The Paddlefish: Status, Management and Propagation

Joe G. Dillard, L. Kim Graham
and Thomas R. Russell
Editors

Barbara K. Bassett
Editorial Assistant and Production Coordinator

Doug Ross, *Artist*

North Central Division
American Fisheries Society
Special Publication Number 7
August 1986

Proceedings of a symposium entitled: "Paddlefish — a threatened resource?" held in conjunction with the 45th Midwest Fish and Wildlife Conference, St. Louis, Missouri, USA, December 6, 1983.

LIBRARY OF CONGRESS CATALOG CARD NUMBER: 86-60218

Copies Available From
Joe G. Dillard
1110 College Avenue
Columbia, Missouri 65201

Printed by Modern Litho-Print Company, Jefferson City, Missouri, USA

Contents

Foreword and Acknowledgments
Joe G. Dillard iv

Guide for Readers vi

Welcome to the Symposium
Charles A. Purkett, Jr. 1

Biology and Life History of the Paddlefish — A Review
Thomas R. Russell 2

Distribution and Abundance of Paddlefish in the United States
Thomas W. Gengerke 22

Threats to Paddlefish Habitat
Rollin D. Sparrowe 36

Effects of Commercial Fishing on Paddlefish Populations
Ronald W. Pasch and C. Michael Alexander 46

Effects of Dams and Other Habitat Alterations on Paddlefish Sport Fisheries
Dennis G. Unkenholz 54

An Overview of Current Management Practices for Paddlefish Fisheries
Allen A. Elser 62

The Role of Regulations in Managing Paddlefish Populations
David L. Combs 68

The Culture of Paddlefish — A Review of Methods
L. Kim Graham, E. Jerry Hamilton, Thomas R. Russell and Charles E. Hicks 78

Establishing and Maintaining Paddlefish Populations by Stocking
L. Kim Graham 95

Opportunities in Paddlefish Aquaculture
Kenneth J. Semmens and William L. Shelton 106

Symposium Wrap-up
The Editors 114

An Indexed Bibliography of the Paddlefish (*Polyodon spathula*)
L. Kim Graham 118

Foreword and Acknowledgments

This symposium grew out of a need for more information about the paddlefish, a species which has recently sparked both public and professional interest. Paddlefish habitat, and in some instances paddlefish populations, have been greatly reduced by water resource development projects over several decades. The range of the species has also diminished. Now a new threat has arisen. Within the last few years the demand for paddlefish eggs for caviar has put a severe strain on various populations. In addition, the mystique of this large, freshwater fish with its unusual features and life habits has made it an object of curiosity.

Our basic objectives in this project were to review, synthesize, present and publish the currently available information on paddlefish, especially information on threats to its survival. We were aware when planning the symposium that much was known but not reported.* Therefore, we invited experts on paddlefish biology, resource managers knowledgeable about threats to habitat, and fisheries managers who had tried various strategies, to review and present what has been done and what needs to be done to better manage the paddlefish resources of the United States.

Paddlefish fisheries in the western, midwestern and southern United States have declined in some areas, but have increased and become more important in others. Papers in this proceedings emphasize that the paddlefish is a unique species with great potential value, and that management agencies must choose either to make the best of a limited resource or to just ignore an important segment of their aquatic resources. Specific management plans for this species, along with public awareness and support, are essential to protect critical habitat, regulate harvest and allow us to answer the question that was posed during the symposium ("Paddlefish — A Threatened Resource?") with a "no" rather than a "yes".

A project of this magnitude is obviously not the work of a few, although Tom, Kim and I nurtured it from 30 April 1982 to date. We especially thank the North Central Division of the American Fisheries Society for their approval, support, funding, patience and understanding in a project that took so long to complete. Thanks also to the speakers/authors who graciously gave of their time and talent. Their excellent presentations were the backbone of the symposium's success.

Dr. Larry A. Jahn (IL) and Mr. Larry R. Mitzner (IA), co-chairmen (1983-85) of the North Central Division's Publication Committee, and Mr. Gary Novinger, chairman (1985-86), performed an invaluable service as reviewers of all the manuscripts. Thanks also to the Missouri Department of Conservation, especially Charles A. Purkett, Jr., Assistant Director, for allowing Tom, Kim and me the time, materials, and support staff necessary to complete such a project.

A special thanks to the Fishery Resources Programs of the U.S. Fish and

*Considerable information included in this proceedings was supported by Federal Aid to Fish Restoration Funds, more commonly known as Dingell-Johnson projects.

Wildlife Service, Department of the Interior for partial funding for the printing and distribution of this proceedings.

Although I am listed as co-editor, Tom, Kim and Barb let me assume the lead role in making the tough decisions that always have to be made in publishing a proceedings such as this. From that perspective, I took tremendous liberties in adapting the presentations for this publication, and in doing so I accept full responsibility for any errors and inconsistencies.

Joe G. Dillard
March 1986

Guide for Readers

We generally followed the format conventions, manuscript components and symbols and abbreviations recommended by the managing editor of the American Fisheries Society. Deviations from these guidelines include the following:

References — Because many authors in this proceedings cite the same literature on paddlefish, we coded all of those references into a comprehensive bibliography that makes up the last chapter of this book. Therefore, a citation such as Purkett[410] or (Purkett[410]) refers to publication number 410 on page 145 in the bibliography. Non-paddlefish references such as Anderson 1975 or (Anderson 1975) are listed at the end of the appropriate paper under ***References*** and are not listed in the comprehensive bibliography.

Measurements — Most measurements are in English units except when experimental values were originally reported in metric units. In these instances, the English equivalent is provided in parentheses.

Fish lengths — Fish lengths are reported as total length (TL). For a further discussion of how paddlefish are measured see page 20.

Abbreviations — All units of measure are abbreviated except for units of time.

Welcome to the Symposium

Charles A. Purkett, Jr.
Missouri Department of Conservation
Jefferson City, Missouri 65102

Welcome to Missouri, to the Midwest Fish and Wildlife Conference and to the Paddlefish Symposium. I'm pleased to see so many of you here, and I'm sure you will benefit from the excellent program. Obviously you will hear the latest information from those people who are currently working in the field.

My involvement with paddlefish goes back to the mid-1950s, to the Osage River in central Missouri where there were large spawning runs of paddlefish. The catch of these fish was very large, especially in 1957, and a year later Paul Barnickol and John Funk asked me to study the fish and the fishery. At that time I knew only three people who were involved with paddlefish: Ted Shields, then working in South Dakota; Fred Meyer, at Iowa State University and Glenn Chambers, who was doing a special study at Central Missouri State University at Warrensburg, Missouri. Previously there had been a few studies, most of them in Illinois. The most notable of these were by L. A. Adams, Dave Thompson, Paul Barnickol and Weldon Larimore.

Hussakof[259] at the 1910 meeting of the American Fisheries Society presented a paper entitled, "The Spoonbill Fishery of the Lower Mississippi." Following the presentation, Dr. B. W. Evermann, who was in the audience, stated: "The queerest thing about the spoonbill is this, that no one has been able to locate definitely its spawning grounds and to find the young fry. Spoonbills five or six inches long are about as small as anyone has ever reported. I think the smallest I've ever seen were eight inches; but fish five to six inches long have been reported." When I saw this it spurred me on to fill this long-standing void, i.e., to find where the fish spawn and observe their eggs and larvae.

The third year of my study I did observe spawning activity, and on April 24, 1960, following a seven-foot drop in the water level, I found what many had been looking for — paddlefish eggs on rocks of gravel bars and hatching young fish in the adjacent water.

Since then a good many workers have made great progress in researching paddlefish biology, and I'll listen with interest to learn what has transpired. I urge you to attend the entire symposium and participate actively in what I know will be excellent discussions of a unique fish.

Biology and Life History of the Paddlefish — A Review

Thomas R. Russell
Fish and Wildlife Research Center
Missouri Department of Conservation
Columbia, Missouri 65201

Abstract

The paddlefish (*Polyodon spathula*) is a primitive species characterized by a predominantly cartilaginous skeleton, a rostrum or "paddle" nearly one-third of the body length, and a smooth skin. Paddlefish are found mostly in large river systems of the Mississippi River Basin and in some rivers entering the Gulf of Mexico. Populations have developed in some large, man-made impoundments, but paddlefish must have access to large, free-flowing rivers to spawn. Their spawning needs include a water temperature near 60 F, clean gravel substrate for egg attachment, and increased water flow to trigger spawning. Because of these exacting requirements, a precise timing of events is necessary to stimulate migration and ensure successful reproduction.

Paddlefish grow relatively rapidly, reach large sizes, and are long-lived. Individuals heavier than 50 lb. are common, and 90-100 lb. specimens are reported frequently. In older established populations, 15 to 20-year-old fish are common, and some may live 30 years. Most males reach sexual maturity at 7-9 years of age; females at 10-12 years.

Paddlefish frequently occur in large groups. They may be found in a variety of habitats, but spend much of their time feeding in quiet or slow-moving water. They feed primarily on zooplankton and aquatic insect larvae, which they filter from the water. Because they are filter feeders, few are caught by conventional fishing methods. In many areas, substantial numbers of paddlefish are caught by blind snagging with large treble hooks.

Paddlefish have adapted to some environmental alterations such as construction of reservoirs on major rivers. However, because of specialized spawning requirements, they have been adversely affected by those alterations that destroy, permanently flood, or prevent access to limited spawning sites. Impacts are not always immediately recognized because paddlefish are long-lived and highly mobile, and their presence is sometimes construed as an indication that the species has not been adversely affected. Wise and practical management requires an understanding of life history requirements of paddlefish and a knowledge of how environmental factors affect this species.

Paddlefish (*Polyodon spathula*) were once abundant in the large rivers in the Mississippi River drainage (Carlson and Bonislawsky[95]). However, within the last 100 years, the peripheral range of the species has shrunk (Gengerke[196]) and there have been significant declines in paddlefish populations in some of our major rivers (Unkenholz[528]). Habitat alteration and destruction, overexploitation, and pollution have detrimentally affected many paddlefish populations (Pflieger[398]; Carlson and Bonislawsky[95]). Water development projects and their operations (Sparrowe[480]) and potential overharvest for roe (Pasch and Alexander[390]) are major current threats to paddlefish populations.

Because of current problems, threats,

and interest in managing the species, research on paddlefish has increased in the last 25 years. Considerable information has been published on various aspects of the biology and life history of paddlefish, but there has been no general review of this material. Vasetskiy[534] reviewed literature on the family Polyodontidae but emphasized systematics, anatomy, and embryonic development. Carlson and Bonislawsky[95] briefly discussed the life history of paddlefish in their review of commercial and sport harvests of the species.

The purposes of this paper are to review the general biology and life history of paddlefish, to relate this information to the species' needs and requirements, and to provide a source of pertinent information that can be quickly located. Information was obtained from a review of the literature and from personal knowledge gained from 20 years of research on the species.

Description

The paddlefish, frequently called spoonbill or spoonbill cat, is one of the largest freshwater fishes, attaining lengths of more than 6 ft. and weights of more than 100 lb. It is the only species of the family Polyodontidae found in North America.

The paddlefish is a primitive species distinguished from other fishes by an elongated rostrum or "paddle" that is approximately one-third the length of the body. The rostrum is absent in newly hatched young, but it starts growing in 1-2 weeks and within 4-5 weeks it is largely developed.

The function of the rostrum is not precisely known. Fishermen generally believe it is used to dig in the bottom for food. Although this theory was supported by some early investigators (Forbes[176]; Imms[261]; Alexander[10]), most believe that the presence of a large number of sensory receptors (Kistler[287]; Nachtrieb[345]) and electro-receptors (Jorgenson[275]) suggests the rostrum is used to detect concentrations of food. The rostrum is not essential for survival. Paddlefish with portions or all of the rostrum missing are in comparable condition to fish of the same size with an intact rostrum. This suggests that the rostrum may not be functionally important in adult fish.

The skeleton of paddlefish is mostly cartilage. The only bony material is primarily in the jaws. The mouth is large, toothless except in young fish, and located under the head at the caudal end of the rostrum. The eyes are small and vision is reportedly poor (Hussakof[259]). Paddlefish have long, thin, close-set gill rakers, which are used to filter food. Like other primitive fishes, the paddlefish has a spiracle and a spiral valve. The body is smooth and appears scaleless, although small patches of minute scales have been found on the pectoral girdle (Wagner[535]), on the upper lobe of the caudal fin (Nikolskii[382]), and between the lateral line and dorsal fin (Collinge[108]). Coloration varies from bluish-gray to almost black above the lateral line, grading to white on the lower sides and belly.

Numerous anatomical studies of paddlefish have been conducted. Vasetskiy[534] reviewed and summarized these, and they are not included in this paper.

In the 1960s and early 1970s there was considerable interest in the immune response system of paddlefish. Although a primitive fish, the paddlefish was considered a key link in phylogenetic development because it possesses a blood serum complement system similar to mammals (Gewurz et al. 1966; Legler[303]), and because antibodies are vigorously formed in response to antigenic stimulation (Clawson[102]). Medical researchers believed that study of the paddlefish immune system might lead to a better understanding of antibody production, so that the tissue rejection phenomenon associated with human organ transplants could be understood and ultimately prevented. However, I am not aware of anyone currently using paddlefish in medical research.

Habitat

The paddlefish is an open water species. Under the natural, unaltered conditions that existed in the late 1800s and early 1900s, this species inhabited the large, free-flowing rivers of the Mississippi Valley. These river systems, with their braided channels, extensive backwater areas and oxbow lakes, provided ideal habitat and supported large paddlefish populations. Adults were usually found in the slower-moving waters of the rivers and were rarely found in small tributaries (Coker[105]). During spring rises, paddlefish moved quickly into bayous and river-lakes. Most of them remained there as long as connection with the river was maintained (Stockard[489]). Some fish stayed in these river-lakes, and moved back into the river during rises. Paddlefish in the bayous and river-lakes were reported to be in better condition than "river fish" (Stockard[489]) presumably because zooplankton, their primary food, was more abundant in the standing waters.

Most of our major rivers have been channelized or dammed in the last 100 years. These alterations have increased water currents, altered natural flow and temperature regimes, and destroyed spawning areas, backwaters, and other preferred habitat. Paddlefish are still present — sometimes in substantial numbers — in some of the altered but still free-flowing rivers, but in general the habitat throughout the range of the paddlefish is now less favorable.

In large, altered streams with strong current, such as the Missouri and Mississippi rivers, paddlefish concentrate in protected natural areas such as pools below sand bars, near physical features like big islands, or near shore irregularities where the current is deflected and velocity reduced (Rosen[425]). They may also gather near man-made structures like dikes, revetments, and bridge supports, where these structures or associated scour holes reduce current (Southall and Hubert[478]). Individuals or small groups of paddlefish may occasionally be found in backwater chutes or nonflowing tributaries, but these habitats appear to be of lesser importance except as feeding areas. Paddlefish seem to prefer water deeper than 4 ft., and during late fall and winter move to areas where the water is 10 ft. or more deep (Rosen[425]). Little is known about habitat preferences in sluggish rivers, but presumably paddlefish would frequent deep pools where food would be most abundant.

Larger paddlefish populations are now usually found in reservoirs which provide optimal feeding environments. Paddlefish spend most of their time in these reservoirs, but the adults migrate into suitable streams in the spring to spawn. Some fish remain in the upstream riverine habitat year-round (Rehwinkel[419]; Russell[450]) but most return to the reservoirs after spawning. Little is known about habitat preference or behavior of paddlefish in standing waters.

Paddlefish are adaptable to a variety of habitats. However, standing or slowly moving waters are preferred during most of the year, are beneficial to most life stages and support greater numbers of paddlefish because of increased food production.

Movement

Paddlefish are known to be mobile; however, very little information is available on precise movements of the species and most is based upon recapture of tagged fish. Although data from tagged fish represent minimum estimates of movement between two points, they do show that paddlefish make extensive upstream and downstream movements.

Most extensive upstream movements of paddlefish are associated with the spring spawning season (Purkett[407]; Elser[151]; Bonislawsky[55]; Rehwinkel[419]; Berg[42]; Southall[477]). Movements of more than 100 mi. are common in the Yellowstone River, Montana (Rehwinkel[419]), and the Osage River, Missouri (unpublished data, 1983, L. Kim Graham, Fisheries Research Biologist, Missouri Department of Conservation, Columbia, Missouri), and movements of more than 145 mi. occur in the Missouri River above Fort Peck Reservoir, Montana (Berg[42]). Some paddlefish have moved more than 200 mi. up the Osage River (Russell[449]). These spawning movements involve large numbers of sexually mature fish.

Dams are barriers to upstream movements, and reservoirs in series isolate and confine paddlefish populations. However, movements within these systems are often extensive. Upstream movement of paddlefish in Garrison Reservoir on the upper Missouri River, North Dakota, is blocked by Fort Peck Dam on the Missouri River and Intake Dam on the Yellowstone River, Montana. However, paddlefish move more than 200 mi., both upstream and downstream, within this system, frequently within a short period of time (Robinson[423]; Elser[151]; Rehwinkel[419]). Robinson[423] reported that three fish moved 200 mi. in 32-45 days, and Rehwinkel[419] reported three paddlefish moved approximately 100 mi. in 17, 35, and 62 days, respectively.

Paddlefish are not confined to reservoirs and frequently emigrate when floodgates are open. Occasionally fish may even go through turbines unharmed although most are killed. Downstream movement of paddlefish from Harry S. Truman Reservoir into Lake of the Ozarks, Missouri, has been significant. Thirty-three of 140 tagged females (23.6%) released in Truman Reservoir were caught in Lake of the Ozarks. They passed through the floodgates, which were opened for extended periods during spring. Four fish passed through both Truman Reservoir and Lake of the Ozarks, which is almost 100 mi. downstream from Truman Dam (unpublished data, 1983, L. Kim Graham, Fisheries Research Biologist, Missouri Department of Conservation, Columbia, Missouri).

Paddlefish frequently move between the pools created by locks and dams on the Mississippi River (Gengerke[195]; Southall and Hubert[478]). Gengerke[195] reported 14 fish moved one to four pools downstream (mean, 57.6 mi.) and four moved one to four pools upstream (mean, 51.5 mi.). Upstream interpool movement was possible because of the navigation locks.

Greatest point-to-point movement has been in open river systems. Unkenholz[526] reported that 12% of the paddlefish tagged below Gavins Point Dam, South Dakota, the lowermost dam on the Missouri River, were harvested outside the South Dakota border. Several fish moved more than 400 mi. downstream, and the farthest movement was 1,200 mi. downstream. Rosen[425] reported that most of the fish he tagged below Gavins Point Dam moved downstream. He reported downstream movement of 42-48 mi. in 3-8 months.

Paddlefish also move considerable distances daily. Transmitter-equipped paddlefish in Lake of the Ozarks made mean overnight (about 12 hours) movements of between 3.4 and 9.5 mi., and maximum movements of 5 to 28 mi. The average 24-hour movement of males was 4.4 mi. (maximum 25 mi.) and of females 3.5 mi. (maximum 27 mi.) (unpublished data, 1983, L. Kim Graham, Fisheries Research Biologist, Missouri Department of Conservation, Columbia, Missouri). These movements were in March, April and May during the spawning period. Movement patterns during other seasons were not determined.

Paddlefish in Pool 13 of the Mississippi River, Iowa, also exhibited greatest movement during the spawning season (Southall and Hubert[478]). During this period they moved an average of 158 ft./hour (0.30 mi./hour) compared to 38 (0.007 mi./hour), 75 (0.014 mi./hour), and 48 ft./hour (0.009 mi./hour) during the prespawning, post spawning, and summer periods, respectively. During the prespawning period paddlefish responded to increased flows with upstream movements, but during other periods movements were not in direct response to changes in water levels.

Long-range movements, both upstream and downstream, seem to be characteristic of the species and may be important to long-term survival and/or population maintenance in some areas. This may be especially significant in areas where dams prevent migration to traditional spawning areas or where reservoirs have flooded existing spawning areas.

Reproduction

Paddlefish do not reach sexual maturity until they are relatively old, and males mature at a younger age than females. Gengerke[195] reported that 6% of the males in the Mississippi River started to mature at 4 years, and all were mature at 9 years of age. In Missouri, most males are sexually mature between 8-10 years of age. When they first reach maturity, males weigh between 15-20 lb.

Females usually do not reach sexual maturity until at least 10 years of age (Gengerke[195]) and many may not mature until they are 12 (Bonislawsky[55]; unpublished data, 1983, L. Kim Graham, Fisheries Research Biologist, Missouri Department of Conservation, Columbia, Missouri). Adams[5] reported females in the Mississippi River were mature at 8-9 years of age. Gengerke[195] reported that 3% of females as young as 6 years were mature, 50% of females at age 10 were mature, and all females 12 years of age and older were mature. Mature females generally weigh at least 30 lb. The smallest sexually mature female collected in Missouri weighed 29 lb. and was 13 years old.

There are no reliable external characteristics for determining sex of paddlefish. Usually, adult males are more slender and in the spring have more tubercles on the top of the head than females, but these characteristics are not reliable. The only precise method to determine sex is by examination of the gonads. Testes lie along the lateral margin of the dorsal surface of the body cavity in a long wavy band. In all but small, immature males the testes lay along extensive fat bodies and are slightly gray. The histology and spermatogenesis of paddlefish were described by Larimore[301].

Ovaries are highly convoluted organs, white when immature, and generally attached to large fat bodies. As ovaries develop they became granular, and small black ova become visible. Later the eggs enlarge, turn gray, and the fat bodies become small. At maturity, the eggs are gray to black and range from 2.0-2.5 mm (0.08-0.10 in.) in diameter (Robinson[423]; Rosen[429]). Fat bodies are usually absent. The histology of ovaries was described by Larimore[301].

Male paddlefish are capable of spawning yearly, but most females are not. The time required to develop mature eggs is not known, but apparently more than 1 year may be necessary. Adult females are frequently collected in spring with ovaries in various stages of development (personal communication, 1983, Dennis G. Unkenholz, Missouri River Coordinator, South Dakota Department of Game, Fish and Parks, Pierre, South Dakota). Elser[151] also reported that females do not spawn annually and Meyer[328] and Sprague[481] concluded that, because of close spacing of annuli on the dentary bone, females may spawn only once every 4 or 5 years.

In Missouri, we also have indirect evidence that mature females make the spawning run only every 2-3 years. Of 140 females tagged and released after their eggs were collected, only three were recaptured the following year during the spawning run. During subsequent years 21 more were recaptured by anglers during the spring spawning season; 11, 6, and 4 were caught 2, 3, and 4 years, respectively, after release. All were gravid when captured.

These observations suggest that most females require 2 years or more to develop mature ova. Similarly, female lake sturgeon (*Acipenser fulvescens*), which mature at 24-26 years of age, spawn once every 3-5 years (Priegel and Wirth 1971). It seems probable, therefore, that fishes which produce large egg masses require more than 1 year to store and mobilize energy for the production of eggs.

The size of the ovaries of paddlefish is not directly proportional to the weight of the female, but larger females usually have larger egg masses. The ovaries of gravid females may comprise 15-25% of the body weight, and females weighing 25-55 lb. will generally contain ovaries weighing 6-8 lb. (Purkett[407]). Stockard[489] reported "lake fish" contained 10-12 lb. of eggs and "river fish" 3-4 lb. of eggs, which suggests that food may affect the number of eggs produced.

Paddlefish produce large numbers of eggs. Eleven fish from the Mississippi River, weighing between 31-52 lb., produced from 148,000-507,000 eggs each. Mean egg production was about 7,700 eggs per pound of body weight (Gengerke[195]). Egg production was higher for paddlefish in the Osage River-Lake of the Ozarks, Missouri, population. Fecundity estimates from 71 females between 31 and 85 lb. ranged from about 184,000 eggs in a 37-lb. fish to slightly over 1,000,000 eggs in a 79-lb. fish. Mean egg production was about 12,000 for each additional pound of body weight (Russell[449]).

Spawning

Early paddlefish researchers were puzzled that no one knew where the species spawned. In fact, considerable effort was devoted to finding young paddlefish. Stockard[489] spent the springs of 1904 and 1905 searching unsuccessfully in the lower Mississippi River for spawning areas and young paddlefish. The smallest paddlefish ever collected by early investigators were seven specimens ranging from 17-

20 mm (0.68-0.80 in.) long, collected from the Mississippi River by Thompson[504]. He concluded that paddlefish spawned in the swift, turbid river channel, and the young remained there. Prior to his collection the smallest paddlefish collected were 35 mm (1.4 in.) long (Barbour[30]) and 74 mm (3.0 in.) long (Danforth[126]). It wasn't until 1960 that precise information on spawning areas was obtained by Purkett[407]. He observed paddlefish spawning over inundated gravel bars in the Osage River, Missouri, during a spring rise. Most of the information about reproduction and early development of young was obtained largely through subsequent investigations in the Osage River (Ballard and Needham[28]; Russell[437, 438, 450]).

Paddlefish require precise timing of events for successful reproduction. They spawn only in flowing water over a gravel substrate. The three most important factors controlling spawning are photoperiod, water temperature and water flow. Within their range, paddlefish may spawn from late March to late June when the water is in the mid-50 F range. Photoperiod and water temperature control the timing of spawning, but an increase in flow is the triggering stimulus. If all three of these conditions are not satisfactory, females resorb their eggs.

Prior to spawning, when the water is near 50 F, paddlefish congregate in deep pools and start moving upstream. As the water warms, their movement increases. When flows are sufficient, paddlefish move quickly to inundated gravel bars and spawn. In the Osage River, paddlefish spawn when flows are sufficient to produce a river stage of 10 ft. and flows of approximately 12,000 cubic feet per second (cfs). Flows of 24,300 cfs in the Mississippi River (Gengerke[195]) and 10,000-20,000 cfs in the Neosho River, Kansas (Bonislawsky[55]), were reported to be associated with movement and spawning of paddlefish.

Paddlefish are very sensitive to changes in water levels, and move downstream quickly into deep pools when water levels start to recede. At the next rise, they move back onto the bars. When a female has completed spawning, she moves quickly downstream.

Because paddlefish spawn when rivers are high and turbid, the spawning act has not been observed. However, considerable activity generally occurs at the water's surface. It is believed that several males accompany a female, and that a spawning "rush" starts near the bottom and culminates at the surface. Splashing can be heard after dark when many fish are spawning (Purkett[407]). Eggs are released during the spawning rush, and immediately after being fertilized they become adhesive and stick tightly to the first object they touch. Most adhere singly to gravel, although some may become attached to twigs or debris. Clean gravel bars are the most favorable substrate and seem to be a prerequisite for spawning.

The maintenance of high water following spawning is important for successful hatching of paddlefish eggs and dispersal of larvae. When water recedes quickly, eggs and larvae are stranded on gravel bars (Purkett[407]). The eggs hatch in about 7 days when the water temperature is in the low 60 F range, so high water levels must remain for at least that long after spawning to ensure a significant hatch. In large streams, rises that stimulate spawning are usually maintained long enough to keep most of the gravel inundated until the eggs hatch. In smaller streams, however, rises frequently recede quickly and many eggs and larvae perish.

Clean gravel and current are important for survival and hatching of eggs. The adhesive coat of the egg holds it in aerated water, and the current prevents siltation. The attachment of the egg to the gravel aids in hatching because the capsule is secure and allows the young to wriggle free. Young in unattached eggs have difficulty freeing themselves from the capsule and many die. After hatching, the young swim up and are immediately swept downstream. A systematic account of the embryonic stages of paddlefish is given by Ballard and Needham[28].

Newly hatched young swim erratically and vertically due to a large yolk sac. This behavior is important because it keeps them in the water column, where currents can carry them from shallow water into areas where food is available. In 3-5 days most of the yolk is absorbed, the mouthparts develop and the young are ready to feed. The young do not have a rostrum or paired fins at this time and do not resemble the adults. After about 1 week they start developing rapidly, and within 10 days to 2 weeks the rostrum begins to develop. At approximately 1 month of age the young are easily recognized as paddlefish.

Feeding

Although early investigators believed that paddlefish used the rostrum to stir up the bottom and feed on dislodged organisms (Forbes[176]; Imms[261]; Alexander[10]), most studies have shown that paddlefish feed primarily on zooplankton (Forbes[176]; Stockard[489]; Eddy and Simer[143]; Barbour[29]; Ruelle and Hudson[432]; Rosen and Hales[427]). Their long, thin gill rakers enable them to efficiently filter zooplankton and other food organisms from the water. Paddlefish have been reported to consume fish (Meyer[328]; Fitz[172]) and at times considerable numbers of insects (Meyer[328]; Hoopes[251]), but these appear to be isolated instances.

Like adults, young paddlefish feed almost exclusively on zooplankton, but their mode of feeding is different because gill rakers are either absent or nonfunctional when the young fish first start to feed. Young paddlefish are particulate feeders, selecting and capturing individual food items (Rosen and Hales[427]; Michaletz[332]). They feed primarily on large cladocerans and occasionally on small insect larvae, like chironomids. Even when cladocerans are not abundant, young paddlefish select them. Copepods are generally not eaten, probably because they are more difficult to capture than the slower cladocerans.

Gill rakers appear on the gill arch when the fish are approximately 3 in. total length (TL — See Appendix A for a discussion on measuring paddlefish) (Michaletz[332]) to 4¾ in. TL (Rosen and Hales[427]), but they are not functional until the fish are approximately 5 in. TL (Michaletz[332]). Rosen and Hales[427] reported that paddlefish were approximately 16-18 in. TL before their gill rakers were fully functional. However, Michaletz[332] reported that, in aquaria, paddlefish approximately 5 in. TL would switch from particulate to filter feeding depending upon the size of food available. When small food items were introduced, young fish fed by filtering, but switched to particulate feeding when larger cladocerans were added. This is an important adaptation and demonstrates that small fish are capable of filtering with less than fully developed gill rakers. When paddlefish become filter feeders, the growth rate generally increases because of increased food consumption (Michaletz[332]).

Once gill rakers are fully developed, paddlefish feed almost exclusively by filtering, and consume organisms roughly in proportion to their abundance, except for small nauplii and rotifers which are consumed in disproportionately small amounts (Eddy and Simer[143]; Rosen and Hales[427]). Rosen and Hales[427] reported paddlefish in the Missouri River below Lewis and Clark Lake, South Dakota, fed mostly in spring and fall and ceased to feed or fed little from late June to early September. They attributed the lack of summer feeding to an unstable environment caused by increased discharge from an upstream reservoir. Adult paddlefish in the Osage River, Missouri, fed actively during the summer and had an average of 140 ml of food in the gut (unpublished data, 1983, L. Kim Graham, Fisheries Research Biologist, Missouri Department of Conservation, Columbia, Missouri). This probably reflected the lack of flow and resulting increased production of food in the Osage River during the summer. The observations of Rosen and Hales[427] suggest that feeding opportunities are reduced in most rapidly flowing waters. These areas are less productive of zooplankton, so paddlefish numbers and growth rates are undoubtedly less than in standing or slowly moving waters.

Age and Growth

Paddlefish are aged by sectioning the dentary bone of the lower jaw and viewing the sections under magnification. Adams[5] used otoliths and dentary bones but reported that otoliths were satisfactory only for aging young fish. Annuli on sections of the dentary bone are frequently difficult to identify, especially in older fish, because of crowding and the presence of halo bands. These two factors make precise aging difficult (Russell[443]), and undoubtedly account for much variation in ages and growth rates reported in the literature.

Paddlefish are long-lived. Fish up to 30 years of age have been reported (Purkett[409]), and paddlefish older than 15 years are common in many populations. Where fisheries are sustained by sexually mature adults, the majority of fish harvested range from 7-18 years of age (Table 1).

Paddlefish grow rapidly compared to most warmwater species, especially during their first year of life. Where food is unlimited, young paddlefish can reach more than 20 in. TL during the first year. However, most grow at a slower rate. Growth in hatchery ponds has ranged from about 1-5 mm/day (0.04-0.20 in./day). In general, paddlefish reach an average total length of 10 in. in less than 100 days (Kallemeyn[277]; Unkenholz[517]; Michaletz[332]; unpublished data, 1983, L. Kim Graham, Fisheries Research Biologist, Missouri Department of Conservation, Columbia, Missouri). In extensively fertilized ponds, paddlefish have grown from fry to about 28 in. TL in about 110 days (personal communication, 1983, E. Jerry Hamilton, Fisheries Area and Hatchery Manager, Missouri Department of Conservation, Sweet Springs, Missouri). Ruelle and Hudson[432] reported young paddlefish grew at the rate of 2.7 mm/day (0.11 in.) in Lewis and Clark Lake, and were about 12 in. TL by September 30. Fastest growth of young-of-the-year paddlefish was reported from Fort Gibson Reservoir, Oklahoma, where they grew at a rate of 0.17 in./day (4.3 mm/day) and averaged 28.4 in. TL by December (Houser and Bross[253]). This exceptional rate of growth was attributed to an abundant plankton population, which developed following a doubling of the surface area of the reservoir and flooding of extensive vegetation. It is unlikely that paddlefish grow at this rate in most natural environments, but it does demonstrate the growth potential of the species when food is not a limiting factor.

Growth usually slows after the first year. By about Age V, paddlefish grow at a rate of approximately 2 in./year, although growth is highly variable (Table 2). The fastest growth reported is from Table Rock Lake, Missouri (Graham[218]). Paddlefish in Oklahoma (Combs[112]) also grow rapidly and at a faster rate than in Old Hickory Reservoir, Tennessee (Pasch[393]), the Mississippi River in Iowa (Gengerke[195]), or the Missouri River in Montana (Robinson[423]; Berg[42]) and South Dakota (Ruelle and Hudson[432]).

Greatest gains in weight occur when growth in length slows. After paddlefish reach about 5 years of age, weight gains are generally rapid, and in the next 5 years they frequently double or triple their weight (Table 2). Growth of paddlefish in reservoirs is faster than in rivers (Table 2), which probably reflects increased abundance and availability of food.

Females generally grow slightly faster in length than males. Once they reach sexual maturity, females usually weigh more than males of the same length or age (Table 3) because of their large ovaries. Swingle[493] reported weight gains from 100-363% for paddlefish in fertilized ponds in Alabama. Fish weighing from 1.5-11.2 lb. when stocked weighed 6.9-26.5 lb., respectively, from 114-144 days later. These rates are extremely fast, much faster than would be expected in natural populations.

Fisheries

Paddlefish provide important fisheries, both sport and commercial, in many areas (Carlson and Bonislawsky[95]). Because paddlefish are filter feeders, few are caught by traditional fishing methods. Most are caught by blind snagging. The gear consists of a stout rod and reel, and heavy line equipped with large treble hooks and a heavy sinker. Anglers fish by trolling or by casting from the bank. The line is retrieved with a jerking motion and the weight keeps the hooks near the bottom where paddlefish prefer to stay. Best fishing success occurs where paddlefish concentrate, usually in deep pools in rivers or below dams that are barriers to upstream movements. Electronic fish locators are becoming popular for finding concentrations of paddlefish.

The two largest paddlefish sport fisheries are in the Osage River, Missouri, and the Yellowstone River, Montana, where the average annual harvests exceed 60,000 lb. (Carlson and Bonislawsky[95]). Catches from these areas are primarily adult fish with an average weight near 40 lb. Important sport fisheries also exist below some of the dams on the upper Missouri River, below some of the locks and dams on the Mississippi River, and below Kentucky Dam in Kentucky (Carlson and Bonislawsky[95]). Sport fisheries have declined in some areas because of lack of recruitment, most notably in the tailwaters of Big Bend and Fort Randall dams, South Dakota (Unkenholz[528]).

Paddlefish are still an important commercial species, although catches are less than at the turn of the century. In 1897 the total commercial harvest was about 2,500,000 lb. (Coker[105]). The average annual harvest for the period 1965-1975 was about 530,000 lb., and approximately 40% of the harvest was in the Tennessee River (Carlson and Bonislawsky[95]). Nearly 75% of the commercial harvest now occurs in the

Table 1. *Age composition of paddlefish harvested from various populations and locations.*

	Location				
Age	Miss. R. Pool 13[1] Iowa	Miss. R. Pool 19[2] Iowa	Mo. R. above Ft. Peck[3] Montana	Mo. R. below Ft. Peck[4] Montana	Yellow-stone R.[5] Montana
1	4	1			
2	11	6			
3	111	11			
4	95	14			
5	131	25			
6	46	73	1		
7	68	20	1		
8	40	17	2	1	
9	16	12	3	3	4
10	12	6	1	1	26
11	7	9	6	4	134
12	5	9	15	4	123
13	3	6	12	4	86
14	4	3	13	7	44
15	2	5	14	5	38
16	4	1	8	4	69
17	2		7	11	179
18	4		6	6	150
19		2	10	11	103
20		1	10	11	21
21			9	8	
22		1	4	6	1
23			5	8	
24			1	6	
25			2	2	
26			1	2	
27				4	
28				1	
29			1		
30					
No. of fish	565	222	132	109	978

[1]Gengerke[195] [3]Berg[42] [5]Rehwinkel[419]
[2]Meyer[328] [4]Robinson[423] [6]Unkenholz[522]

Table 1, *continued.*

Mo. R. below Lewis & Clark[6] South Dakota	Osage R.[7] Missouri	Osage R.[8] Missouri	Table Rock Lake[9] Missouri	Grand Lake[10] Oklahoma	Old Hickory[11] Tennessee
12		2			3
3	2	6			7
14	14	1	1		4
15	7	13	7		5
14		12	30		4
10	1	3	20	13	6
14		5	27	16	8
7	1	6	26	24	1
8		7	26	18	9
	1	6	29	15	4
3	4	8	42	14	7
2	3	26	26	8	4
	5	11	12	3	
1	7	40			3
	7	47			2
	10	47			
	26	84			
	19	58			
	21	44			
	6	39			
	11	26			
	2	19			
	3	16			
	2	10			
	2				
	1				
	1				
	1				
	1				
103	158	536	246	111	67

[7]Purkett[409]
[8]Russell[443]
[9]Graham (unpublished data)
[10]Combs[112]
[11]Pasch[393]

Table 2. *Mean total length in inches, and weight in pounds, of paddlefish from various populations. Total lengths were converted when necessary using appropriate regressions listed in Appendix A, Table 1.*

Location

	Table Rock[1] Missouri		Lake Ozarks[1] Missouri		Grand Lake[2] Oklahoma		Old Hickory[3] Tennessee		Yellowstone R. at Intake[4] Montana	
Age	in.	lb.	in.	lb.	in.	lb.	in.	lb.	in.	lb.
1							17.9			
2			26.8	2.5			27.4			
3							37.0			
4	50.2	21.7	35.5	6.0			32.2		37.8	
5	53.5	25.2	45.0	12.7			38.7		38.2	
6	56.4	30.4	48.0	16.6	52.0	24.0	40.3		45.2	
7	58.7	32.3	52.8	23.0	53.6	25.3	47.3		43.6	
8	63.3	40.7	53.8	21.6	56.9	36.5	53.0		48.0	
9	64.1	39.4	56.2	24.6	59.3	41.8	50.4		48.9	
10	65.5	43.2	58.8	30.8	62.4	50.8	50.7		50.4	
11	69.7	53.0	61.1	33.3	63.5	56.3	52.1		52.2	
12	70.4	58.1	64.7	41.7	64.7	60.3	57.6		54.3	
13	70.3	57.8	65.8	45.3	66.8	69.3			55.2	
14			66.3	47.3			60.6		48.0	
15			68.5	54.3			56.9		51.9	
16			70.8	57.2					53.6	
17			74.6	62.8					54.1	
18			69.8	76.3					55.6	
19									54.5	
20			75.1	87.0					58.1	
21										
22										
23									57.5	
24										
25									57.4	
26										
27										
28										
29										
No. of fish	245		282		111		67		343	

[1]Graham (unpublished data)
[2]Combs[113]
[3]Pasch[393]
[4]Rehwinkel[419]

Table 2, *continued.*

Mo. R. above Ft. Peck[5] Montana		Mo. R. below Gavins Pt.[6] S. Dakota		Miss. R. Pool 19[7] Iowa		Miss. R. Pool 13[8] Iowa	
in.	lb.	in.	lb.	in.	lb.	in.	lb.
				8.1		24.2	1.0
				21.8	1.6	26.0	1.3
				27.9	3.3	33.7	3.9
		41.8		34.5	4.5	37.2	5.8
		42.2		36.0	5.6	39.9	7.7
40.1		45.2		38.1	6.3	42.7	10.2
48.0		46.4		39.5	7.4	44.4	11.4
52.7		58.9		40.4	7.9	47.3	16.3
53.5		46.8		42.0	9.1	49.7	19.6
53.9				42.9	12.3	51.0	22.6
55.8		54.3		43.6	11.7	51.8	27.2
55.9		55.1		45.2	11.9	53.6	32.3
58.0				46.3	12.3	55.0	40.3
60.3				47.3	13.5	57.6	33.2
61.8				48.2	16.3	59.5	44.9
59.6				48.1	19.5	59.7	42.4
65.8						61.1	47.9
64.2						62.9	50.4
65.7				51.0	21.0		
64.7				52.0	25.0		
66.6							
68.5							
68.1							
69.7							
70.9							
71.2							
74.0							
131		102		221		565	

[5]Berg[42]
[6]Unkenholz[522]
[7]Meyer[328]
[8]Gengerke[195]

Table 3. *Mean total length in inches, and weight in pounds, of male and female paddlefish from selected populations. Total lengths for Missouri were converted using the conversion factor listed in Appendix A, Table 1.*

	Males							
	Table Rock[1] Missouri		Lake Ozarks[1] Missouri		Mo. River[2] Montana		Yellowstone R.[3] Montana	
Age	in.	lb.	in.	lb.	in.	lb.	in.	lb.
1								
2			26.3	2.5				
3								
4	50.2	21.7	34.8	6.0				
5	53.5	25.2	44.2	12.7				
6	56.4	30.4	47.1	16.6	40.2			
7	57.8	31.0	51.8	23.0	48.0			
8	62.6	37.4	52.8	21.6	52.8			
9	63.3	37.2	54.3	23.4	53.5		46.2	
10	64.5	38.7	56.4	27.9	53.9		48.2	
11	63.7	37.4	57.5	28.2	54.7		49.6	
12	64.8	41.0	58.8	32.8	55.9		51.0	
13			59.3	34.6	56.3		52.3	
14			61.7	36.3	57.1		53.3	
15			60.9	45.5	57.9		53.6	
16					58.7		54.6	
17					59.1		54.8	
18					58.7		56.3	
19					59.1		54.5	
20					58.7			
21					61.0			
22								
23								
24								
25					63.0			
26								
27								
28								
29								
No. of fish	138		138		69		500	

[1]Graham (unpublished data) [2]Berg[42] [3]Rehwinkel[419]

Table 3, *continued.*

Females							
Table Rock[1] Missouri		Lake Ozarks[1] Missouri		Mo. River[2] Montana		Yellowstone R.[3] Montana	
in.	lb.	in.	lb.	in.	lb.	in.	lb.
64.1	39.5						
64.7	46.9						
66.6	46.7						
66.4	47.0	62.5	40.8				
70.9	56.0	66.6	46.4	61.4			
70.6	58.8	66.3	46.8				
70.3	57.8	66.7	49.6	61.4			
		66.1	50.7	65.4			
		67.7	55.0	66.9		58.5	
		69.5	57.2	66.5		58.7	
		73.2	62.8	66.9		60.8	
		68.5	76.3	66.9		62.6	
				67.3		63.8	
		73.7	87.0	67.3		63.9	
				67.3			
				68.5			
				68.1			
				69.7			
				70.9			
				71.3			
				74.0			
107		143		53		477	

Tennessee, Mississippi, Arkansas and Cumberland rivers.

Paddlefish eggs have been used for caviar, and in the last 5 years there has been increased demand for eggs for this purpose. Reports of paddlefish eggs selling for $30-$40/lb. have stimulated interest in exploiting the resource in some areas. There is concern that populations may be overharvested, and several states have implemented special regulations to protect the paddlefish resource (Pasch and Alexander[390]).

Discussion

Paddlefish have been the subject of many investigations since the late 1800s. Within the last 20-25 years the emphasis has shifted to management, because of environmental changes that have adversely affected this resource and because of specific problems, such as high harvest, that require attention. Some localized problems are being managed by regulations. Others, such as threats to the population from environmental alterations, are more widespread and require special consideration and management.

Some paddlefish populations and fisheries have declined substantially because of destruction of spawning areas. Others may be threatened by construction or operation of watershed projects (Unkenholz[528]; Sparrowe[480]). Although other factors can affect the abundance of paddlefish, the destruction of spawning areas should be of primary concern because the effects are permanent.

Because of the paddlefish's precise spawning requirements, spawning habitat must be preserved to maintain self-sustaining paddlefish populations. Surprisingly, no spawning sites have been identified except in the Osage River, Missouri (Purkett[407]; Russell[450]), although the spawning requirements of the species are known. Paddlefish fry have been collected in several river systems (Mississippi River — Pool 13, Helms[244]; Missouri River above Lewis and Clark Lake, Unkenholz[526]; Missouri River below Lewis and Clark Lake, Kallemeyn and Novotny[280], Tondreau[510]; Cumberland River below Cordell Hull Lake, Pasch[393]; Neosho River above Grand Lake of the Cherokees, Combs[112]; Missouri River above Fort Peck Reservoir, Berg[42]). These limited collections indicate that spawning sites may be scarce.

Many paddlefish populations contain young and immature fish, but that does not necessarily mean that the population is reproducing. Because paddlefish move long distances, the young could be recruited into populations where reproduction is limited or nonexistent from a spawning population many miles away. Information in Missouri suggests that some paddlefish hatched in the Osage River above Lake of the Ozarks were recruited into the population of the lower Osage River below Lake of the Ozarks. Within 1 month after hatching, trawl samples showed downstream dispersal and concentrations of young paddlefish in the lower portion of Lake of the Ozarks, more than 100 mi. from where they were hatched. During years when water releases through the floodgates were high, many small paddlefish probably passed through the reservoir into the lower river. Adult paddlefish regularly pass through these floodgates, and it is probable that the young, especially those less than 2 months old, would be more vulnerable to the strong currents. Additional support for this belief is the fact that the majority of the paddlefish harvested in the tailwaters below Lake of the Ozarks are immature, with an average weight of less than 12 lb. (Hanson[236]).

A similar situation exists in South Dakota. Young-of-the-year paddlefish have been collected annually in Lewis and Clark Lake, the most downstream reservoir on the Missouri River, since 1965 (Unkenholz[526]). The adult population in the lake is small and the fishery is insignificant (Unkenholz[528]). However, an important fishery consisting largely of immature paddlefish exists in the tailwaters of Lewis and Clark Lake (Unkenholz[527]). Although limited reproduction has been documented in the Missouri River downstream from Lewis and Clark Lake (Kallemeyn and Novotny[280]), the tailwater fishery is probably maintained or substantially augmented by recruitment of young paddlefish from the reservoir during high discharges. This belief is further supported by the fact that Walburg[539] documented larval paddlefish in the reservoir discharge from Lewis and Clark Lake. It is likely that significant numbers of young paddlefish leave the reservoir during late spring and summer, when discharges are sufficient to equal release of the storage capacity of the reservoir in 4-10 days (personal communication, 1983, Dennis G. Unkenholz, Missouri River Coordinator, South Dakota Department of Game, Fish and Parks, Pierre, South Dakota). Extensive downstream movement of paddlefish from below Lewis and Clark Lake has also been documented (Unkenholz[527]), and it is probable that fish from that area contribute substantially to the paddlefish population in the lower Missouri River.

Circumstantial and indirect evidence suggests that reproduction in localized areas — such as the Missouri River above Lewis and Clark Lake and the Osage River above Lake of the Ozarks — provides young fish to maintain some paddlefish populations that are isolated from spawning sites. Of great concern is that destruction of these spawning areas would adversely affect not only the viability of the local spawning populations, but also other paddlefish populations dependent upon recruitment from these areas.

An important step to management of paddlefish populations is to determine where the populations spawn so that those areas can be protected. This information, plus an understanding of the needs of the species, is essential in formulating the best management strategies for paddlefish populations and in evaluating the effects of proposed watershed and water resource development projects on the species.

References

GEWURZ, H., J. FINSTAD, L. H. MUSCHEL, AND R. A. GOOD. 1966. Phylogenetic inquiry into the origins of the complement system. Pages 105-117 *in* R. T. Smith, P. A. Miescher and R. A. Good, editors. Phylogeny of immunity. University of Florida Press, Gainesville, Florida, USA. 276 pages.

PRIEGEL, G. R. AND T. L. WIRTH. 1971. The lake sturgeon: Its life history, ecology, and management. Publication 240-70. Wisconsin Department of Natural Resources, Madison, Wisconsin, USA. 19 pages.

Appendix A. Measuring Paddlefish

Fishery workers have traditionally measured paddlefish by total length (TL). Typically this measurement was from the tip of the rostrum to the upper tip of the caudal fin. Because of the unusual anatomy of the fish, total length measurements are not always taken the same way. Also, because some fish have rostrums or caudal fins that are damaged, frayed or partially missing, an accurate total length measurement is often not possible.

Because of these problems, Ruelle and Hudson[432] proposed that body length (BL), as measured from the anterior edge of the eye to the fork of the tail, be the standard measurement for paddlefish. Most investigators have since reported paddlefish length as body length, and regressions and conversion factors have been calculated for some populations to permit comparisons with earlier work (Table 1). Unfortunately there is no accurate way to convert total length measurements from an early study to body length measurements using reported regressions because the regressions were calculated for a specific population and method of measuring.

In my manuscript I report all lengths as total length to attain as much consistency and accuracy as possible and to permit comparisons of lengths of fish over a broad time span. My reasons for not converting to body lengths are: 1) all measurements of paddlefish prior to 1977 were reported as total lengths, 2) the regressions and conversions calculated were for specific populations, 3) the regressions were only for adult paddlefish (25 in. TL and longer), 4) most recent investigators reported both body length and total length or regressions to calculate it, and 5) almost all lengths of small paddlefish in the literature are total lengths. I calculated a regression for small paddlefish (82-291 mm TL) and used it to convert the body lengths of small fish reported by Rosen and Hales[427]. I realize the conversions may not be accurate but hope they are close approximations for the ranges I use.

Table 1. *Regression and conversion factors reported for converting body length (BL) of paddlefish to total length (TL).*

Location	Regression/conversion	Length range (TL) of fish used in calculations
Mississippi River, Iowa[1]	$TL = 225 + 1.198\ BL$	635-1,524 mm
Lake Oahe, South Dakota[2]	$TL = 200.61 + 1.286\ BL$	890-1,505 mm
Old Hickory Reservoir, Tennessee[3]	$TL = 2.22\ BL^{0.93}$	345-1,577 mm
Osage River, Missouri[4]	$TL = 1.62\ BL$	31-67 in.
Missouri (hatchery fish)[5]	$TL = -1.08 + 1.8565\ BL$	82-291 mm

[1]Gengerke[195]
[2]Ruelle and Hudson[432]
[3]Pasch[393]
[4]Graham (unpublished data)
[5]Russell (unpublished data)

Distribution and Abundance of Paddlefish in the United States

Thomas W. Gengerke
Iowa Conservation Commission
Spirit Lake, Iowa 51360

Abstract

Significant declines of principal paddlefish (*Polyodon spathula*) stocks have occurred in the Mississippi, Missouri, Ohio and Red rivers since the turn of the century, while notable increases have been documented in the Tennessee, Cumberland and Arkansas rivers. Peripheral range has dwindled; however, paddlefish still occur over most of their historic range, and in some instances appear to be reinvading previously lost habitat. Paddlefish still occur in 22 states. The total commercial harvest has decreased by approximately 60% since the early 1900s, with individual fisheries having decreased by as much as 95%. Eleven states allow commercial harvest. Commercial fisheries in the Tennessee, Cumberland and Arkansas rivers now account for almost 70% of the reported commercial catch. Sport harvest has increased and sport fisheries presently exist in 17 states. The average harvest from self-sustaining sport fisheries, based upon creel surveys conducted from 1973 through 1979, was equal to about 70% of the total commercial catch reported for 1975. The largest paddlefish fisheries presently exist in the Tennessee, Mississippi, Cumberland, Arkansas, Yellowstone and Osage rivers. Paddlefish are protected from all forms of exploitation in four states and are extirpated in four states, all of which are located on the present day peripheral range.

The distribution and abundance of paddlefish (*Polyodon spathula*) populations and fisheries in the United States have changed markedly since the turn of the century. The purposes of this paper are (1) to illustrate historical changes that have occurred in the distribution of paddlefish, (2) to define the current range of paddlefish, (3) to provide perspectives on changes which have occurred and in many instances are still occurring, and (4) to offer some insight into the current status of paddlefish populations in the United States. Information for this paper was obtained from a literature review and from personal communication with recognized experts on paddlefish fisheries throughout the United States. A questionnaire (Appendix A) to determine paddlefish occurrence, abundance, status and classification, and the impacts of habitat changes on paddlefish sport fisheries, was mailed to resource agencies throughout the historical range of the species. Telephone contacts (Appendix B) ensured a complete response to this inquiry.

Distribution

Paddlefish were historically abundant in large streams of the Mississippi River and adjacent Gulf Slope drainages (Burr[86]; Carlson and Bonislawsky[95]; and Pflieger[398]). Specimens occurred in Lake Erie (Trautman[512]) prior to 1903 and in other Great Lakes sometime around the turn of the century (Hubbs and Lagler[256]; Halkett[232]; McAllister[318]); however, these relict populations have apparently been lost (Eddy and Underhill[144]).

Populations have also been extirpated from areas within Iowa, notably Okoboji and Spirit lakes and the upper reaches of the Des Moines River; from Nebraska, where the far western reaches of the Platte River no longer contain paddlefish; from the Allegheny and Clarion rivers in Pennsylvania; from Tennessee, at Reelfoot, Dale Hollow and J.P. Priest lakes; from New York, where a single specimen was taken from Lake Chautauqua; and from North Carolina, where Professor Edward Drinker Cope found paddlefish in the French Broad River during surveys conducted prior to 1867. Present range appears to be dwindling in the Galveston Bay, Sabine Lake and Trinity River areas within Texas, and paddlefish are no longer observed above the Prairie du Sac Dam on the Wisconsin River in Wisconsin. Paddlefish were known to exist in Ontario (Burr[86]) but have since been extirpated from Canada.

Peripheral range for paddlefish has diminished and they are considered abundant only in portions of the Mississippi River drainage (see map, inside front cover). Paddlefish are found in the Mobile Bay-Alabama River drainage (Carlson and Bonislawsky[95]); however, they are not considered abundant and are no longer found above Thurlow Dam on the Tallapoosa River nor are they found above Jordan Dam on the Coosa River (personal communication, 1983, William Reeves, Assistant Fisheries Chief, Department of Conservation and Natural Resources, Montgomery, Alabama). With few exceptions, the reduction in range has been confined to peripheral reaches of the historical distribution.

Indices of Abundance

Sport and commercial harvests of paddlefish stocks are difficult to compare or analyze from a continent-wide perspective due to absence of data and/or biases inherent in the collection of such data. Sport fisheries have been assessed in 10 areas. Mean annual harvest from these fisheries is approximately 336 t. (Table 1). Substantial sport fisheries exist in other areas, but have not been investigated. In contrast, commercial harvests are reported for virtually all areas supporting a commercial fishery. Commercial catch statistics are usually considered to represent minimum estimates and may be valid only for trend analysis. Present day commercial harvest of paddlefish in the United States is approximately 473 t. per year (Table 2).

Abundance as Indicated by Commercial Harvest

Commercial harvest of paddlefish started around 1900 after lake sturgeon (*Acipenser fulvescens*) stocks were depleted and demand for an alternative caviar source escalated (Carlson and Bonislawsky[95]). Early catch statistics (Table 2) as reported by Carlson and Bonislawsky[95] were scattered and many times incomplete. Fishery statistics records since 1950 have been complete and accurate, except for 1951-1954, 1971 and 1974.

The largest reported harvest of paddlefish from the Mississippi River was 938 t. in 1899. During the next 23 years reported harvest declined by 70%, and by 1931 harvest had dropped to 108 t., a 90% decline from 1899. Although there have been numerous fluctuations in harvest since 1931, the catch has never approached that reported in 1899. By 1975 commercial harvest of paddlefish from the Mississippi River had declined approximately 95% from that reported around 1900.

Table 1. *Number, mean size and average harvest of paddlefish from selected sport fisheries, 1959-1979.*[1]

		Number				
River	Location	1959	1961	1962	1963	1964
Cumberland	Barkley Dam, KY					
Mississippi	Lock and Dam 12, IA					
Missouri	Hwy. Bridge 171, MT					
Missouri	Big Bend Dam, SD					
Missouri	Fort Randall Dam, SD	4,491	448		229	
Missouri	Gavins Point Dam, SD					
Missouri	Oahe Dam, SD		3,200	4,434	700	
Osage	Warsaw, MO	8,210	4,050	755	2,170	5,320
Tennessee	Kentucky Dam, KY					
Yellowstone	Intake, MT					2,562
Total						

[1]Data from Carlson and Bonislawsky[95].

This decline was largely due to changes in the river caused by channelization and construction of locks and dams. These changes produced two distinct river environments: the impounded, pooled portion upstream from St. Louis, and the open river portion downstream from St. Louis.

Commercial harvest declined in the impounded portion until the mid-1950s, after which it increased, though never to previous levels (Table 3). Harvest in the lower pools remained relatively high, but commercial catch in the uppermost pools never recovered. Harvest from the open river portion has declined since the 1960s. Approximately 90% of the total commercial harvest from the Mississippi River now comes from the impounded portion. Records of commercial harvest also show severe declines since the 1800s from the Ohio, Missouri and Mobile rivers (Carlson and Bonislawsky[95]).

The largest commercial fishery for paddlefish now occurs in the Tennessee River. Prior to 1931 a rather insignificant harvest was reported for this river; however, by 1950 harvest had increased from 4 t./year (1931) to over 102 t./year (1950). By 1975, the Tennessee commercial paddlefish fishery accounted for over 45% (218 t.) of the entire commercial harvest (475 t.) in the United States. Harvest in the Cumberland and Arkansas rivers increased from 22 t. (1899) to 110 t. (1975), and now collectively represents over 20% of the commercial harvest in the United States.

The Mobile Bay-Alabama River drainage is the only system independent of the Mississippi River drainage where paddlefish are now found (Carlson and Bonislawsky[95]). The commercial harvest from these areas is considered small (personal communication, 1983, William C. Reeves. Assistant Fisheries Chief, Department of Conservation and Natural Resources, Montgomery, Alabama). However, the opportunity commercial fishermen in this area now have to fish different drainages and the resultant effect upon market reports makes quantitative assessment of this fishery difficult.

Table 1, *continued.*

				Number					
1971	1973	1974	1975	1976	1977	1978	1979	Mean[2] size (lb.)	Average[2] harvest (t.)
						1,294	2,276	7	7
				1,320				13	9
	467	637	385		666			51	13
2,831	1,335	919		600	250	125	132	44	12
Few				Few			Few	61	
	4,293	2,294						13	16
Few								13	
	1,899	2,017	2,213	3,636	1,654			31	35
						102,837	15,050	7	202
	2,675	2,182	1,473	1,410	1,410	2,887	1,727	42	42
									336

[2]Average of all values available for 1973-1979.

Abundance as Indicated by Sport Harvest

Sport harvest occurs throughout most of the existing range of paddlefish. The largest sport fisheries, supported by self-sustaining or augmented populations, exist only in the upper and central portions of the United States (See map, inside front cover). Most of these developed in rivers following reservoir construction and are located immediately above or below a dam (Carlson and Bonislawsky[95]). A number of smaller fisheries occur below lowhead dams. The importance of these is dependent upon whether or not fish are blocked from further upstream movement. The reported mean annual sport harvest from those

Table 2. *Commercial harvest[1] in tons of paddlefish in the United States, 1894-1975.*

River	1894	1899	1922	1931	1950	1960	1970	1975
Ohio	56	59	2	46	30	43	16	8
Mississippi	191	938	253	108	219	35	72	60
Missouri	64	50	16	<1	4	28	4	4
Red and tributaries	0	<1	236	130	107	10	11	14
Arkansas	0	18	9	4	6	7	37	51
Cumberland	0	4	1	1	<1	7	13	59
Tennessee	<1	<1	<1	4	102	150	98	218
Other	206	147	180	182	13	12	51	59
Total	517	1,216	697	475	481	292	302	473

[1]Harvest values are from Smith[469], Townsend[511], Sette[457], Coker[106], Fiedler[166], and Fisheries Statistics of the United States as reported by Carlson and Bonislawsky[95].

Table 3. *Commercial harvest in tons of paddlefish from the impounded portions of the Mississippi River above St. Louis, Missouri, 1894-1980.*

State	1894	1899	1922	1931	1950	1960	1970	1975	1980
Iowa	22	10	24	4		6	20	16	15
Illinois	58	74	40	12	21	14	44	36	26
Missouri	22	54	8	18	2	4	4	6	11
Minnesota	6								
Wisconsin	8		14						
Total	116	138	86	34	23	24	68	58	52

fisheries which have been monitored is 336 t. (Table 1). Important sport fisheries are located on the Tennessee River at Kentucky Dam in Kentucky; the Yellowstone River at Intake, Montana; the Osage River at Warsaw, Missouri; Gavins Point Dam on the Missouri River in South Dakota; Highway Bridge 171 on the Missouri River in Montana; and Big Bend Dam on the Missouri River in South Dakota.

Status by States

All states which were known or suspected to have paddlefish were first sent a questionnaire (Appendix A) and later contacted by telephone (Appendix B) for information pertaining to distribution, abundance and overall perspectives on the status of present day paddlefish populations (Table 4). Of the 26 states where paddlefish occurred, 22 now have populations. Although existence of paddlefish in North Carolina is unknown, there was an unconfirmed report of a single specimen taken in 1983. Populations are considered by resource agency personnel to be stable or increasing in 14 states, declining in 7 and unknown in Oklahoma (Fig. 1). Paddlefish are considered extirpated from Maryland, New York, North Carolina and Pennsylvania.

Alabama (sport and commercial fisheries, classified as commercial, status declining) — Paddlefish occur in the Alabama, Tallapoosa, Coosa and Tennessee rivers. Populations are declining in the upper Alabama River, especially below Millers Ferry Dam. They no longer exist above Thurlow Dam on the Tallapoosa River nor above Jordan Dam on the Coosa River. Distribution of paddlefish in Alabama has not changed appreciably (personal communication, 1983, William C. Reeves, Assistant Fisheries Chief, Department of Conservation and Natural Resources, Montgomery, Alabama). The commercial harvest is greatest in the Tennessee River, where it has increased from just a measurable amount in 1899 to more than 166 t. in 1980.

Arkansas (sport and commercial fisheries, classified as commercial, status increasing) — Paddlefish occur throughout Arkansas and are found in all major river systems including the Arkansas, Mississippi, St. Francis, White, Red and Ouachita rivers. They are most abundant in large rivers and are found in many tributaries. Distribution of paddlefish in Arkansas has not changed, and the most recent information indicates that paddlefish populations are actually increasing statewide. Commercial harvest has

Table 4. *Classification, population status, and type of fishery allowed for all states containing paddlefish.*[1]

State	Classification	Status	Sport fishery	Commercial fishery
Alabama	Commercial	Decline	Yes	Yes
Arkansas	Commercial	Increase	Yes	Yes
Illinois	Commercial	Decline	Yes	Yes
Indiana	Sport	Stable	Yes[2]	No
Iowa	Commercial	Stable	Yes	Yes
Kansas	Sport	Decline	Yes	No
Kentucky	Commercial	Stable	Yes	Yes
Louisiana	Commercial	Stable	No	Yes
Maryland	Threatened	Extirpated	No	No
Minnesota	Protected	Stable	No	No
Mississippi	Commercial	Stable-Increase	Yes	Yes
Missouri	Game	Stable	Yes	Yes
Montana	Sport[3]	Stable	Yes	No
Nebraska	Sport	Stable	Yes	No
New York	Extirpated	Extirpated	No	No
North Carolina	Not Classified	Extirpated[4]	No	No
North Dakota	Commercial	Stable-Increase	Yes	No
Ohio	Endangered	Decline	No	No
Oklahoma	Commercial	Unknown	Yes	Yes
Pennsylvania	Extirpated	Extirpated	No	No
South Dakota	Sport	Decline	Yes	No
Tennessee	Commercial	Stable	Yes	Yes
Texas	Endangered	Decline	No	No
Virginia	Non-game[5]	Stable	Yes	Yes
West Virginia	Sport[6]	Decline	Yes	No
Wisconsin	Watch List	Stable	No	No

[1]Paddlefish are considered to be extirpated from Maryland, New York, North Carolina and Pennsylvania.
[2]Paddlefish may be taken by hook and line. Snagging is not a legal method of take.
[3]Paddlefish are designated as a species of special concern in Montana.
[4]There was an unconfirmed report of a paddlefish being taken the week of October 10, 1983.
[5]Non-game fish are not regulated in Virginia. Fish can be kept if taken by either sport or commercial fishing. Paddlefish are seldom harvested by either method.
[6]Paddlefish are classified as a sport fish in West Virginia, but they are also listed as threatened. Fish can be kept if caught by hook and line.

increased from 58 t. in 1975-1976 to more than 439 t. in 1981-1982, a nearly 700% increase. Sport fisheries occur below Dams 1 and 2 and below the Dardanelle Dam on the Arkansas River (personal communication, 1983, Mike Freeze, Special Projects Coordinator, Arkansas Game and Fish Commission, Lonoke, Arkansas).

Illinois (sport and commercial fisheries, classified as commercial, status declining) — Paddlefish are taken regularly from all major rivers in Illinois but they are much less common now than previously (Smith[471]). Distribution on the Illinois River had already diminished by the early 1900s. Forbes and Richardson[180] as reported by Smith[471] noted that they were "sparingly" represented in their collection and that they had already become rare in the Illinois River. Commercial harvest has been reported from the Illinois, Mississippi, Little Wabash, Sangamon, Kaskaskia and Big Muddy rivers. Populations in these areas have declined during this century. Catch statistics reveal drastic declines in population during the early 1960s followed by a gradual increase until 1978, at which time harvest again began to decline (personal communication, 1983, Dan Sallee, Fisheries Biologist, Illinois Department of Conservation, Aledo, Illinois). Sport fisheries exist on the Mississippi River below Lock and Dam 12, 15, 17 and 19; on the Kaskaskia River below the first dam at Carlyle; and on the Rock River at the Milan and Sterling dams (Carlson and Bonislawsky[95]).

Indiana (sport fishery only, classified as sport, status stable) — Paddlefish are found in the Wabash and White River systems. In the past they were also reported from the Ohio River and from Manitou, Tippecanoe and Maxinkuctee lakes (Blatchley[50]). Indiana Department of Natural Resources personnel believe the distribution of paddlefish has not changed and consider the present population to be stable. Commercial fishing is not permitted, and although paddlefish may be taken by hook and line, there is no viable sport fishery (personal communication, 1983, Joe Janisch, Fishery Staff Specialist, Indiana Department of Natural Resources, Indianapolis, Indiana).

Iowa (sport and commercial fisheries, classified as commercial, status stable) — Paddlefish are indigenous to Iowa and occur throughout the Mississippi and Missouri rivers and lower reaches of the Cedar, Des Moines and Iowa rivers. They were formerly rather common but are now absent from Okoboji and Spirit lakes. The largest paddlefish of record was speared in Okoboji Lake in 1916 (Nichols[379]); it weighed over 198 lb. (Harlan and Speaker[238]). Paddlefish are most abundant in the Mississippi and Missouri rivers. The largest population is in Pool 13 of the Mississippi River (Gengerke[195]). The largest commercial harvest occurs in downstream pools (unpublished data, Iowa Conservation Commission, Des Moines, Iowa). The main sport fisheries occur in the Mississippi River.

Kansas (sport fishery only, classified as sport, status declining) — Paddlefish are not abundant in Kansas, and their distribution has not changed substantially (personal communication, 1983, Bruce Taggart, Regional Fisheries Supervisor, Kansas Fish and Game Commission, Chanute, Kansas). They occur in the Missouri, Marais des Cygnes and Neosho rivers. Paddlefish may also be found in lower reaches of the Kansas River below Bostwick Dam, in the Wakarusa River below Clinton Reservoir, in the Arkansas River, and occasionally in the Ninnescah River in Sedgewick County (Bonislawsky[55]).

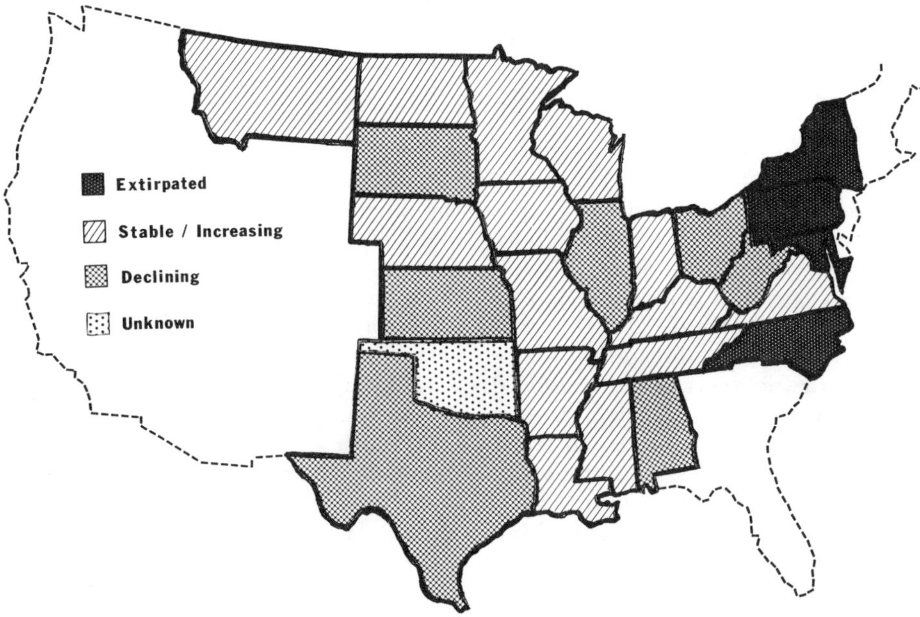

Figure 1. *Status of paddlefish stocks in the United States.*

Paddlefish have not been observed in the Verdigris River since impoundment of the Oologah Reservoir in Oklahoma (Bonislawsky[55]). Construction of Kaw Reservoir in 1975, on the main stem of the Arkansas River in Oklahoma, prevents upstream movement of paddlefish, and it is not known if they still persist in the Arkansas River in Kansas. The largest sport fishery occurs in the Neosho River at Chetopa (Bonislawsky[55]).

Kentucky (sport and commercial fisheries, classified as commercial, status stable) — Paddlefish inhabit the Tennessee, Cumberland, Green, Kentucky, Little Sandy and Ohio rivers and are probably present in all major streams of the state (Clay[103]). They have been reported from the Cumberland River drainage by Jordan and Brayton[271], Kirsh[284], and Turner[514]; from the Ohio River at the Falls by Call[91]; from the Triplett Creek-Licking River drainage by Barbour[29]; from the Wolf River and Ill-Will Creek in Clinton County (Kirsh[284]); and from the lower Kentucky River by Carter[98]. Present distribution is much the same as that reported in 1878 (personal communication, 1983, Ted Crowell, Assistant Director of Fisheries, Kentucky Department of Fish and Wildlife Resources, Frankfort, Kentucky). Paddlefish populations appear to be stable over most of the range, and the population in the upper Cumberland River supports an important commercial fishery.

Louisiana (commercial fishery only, classified as commercial, status stable) — Paddlefish occur statewide in the Mississippi, Red, Ouachita, Amite and Pearl rivers. Distribution remains unchanged (personal communication, 1983, Richard Brassette, Fisheries

Biologist, Department of Wildlife and Fisheries, Baton Rouge, Louisiana), and based upon commercial harvest records, the present populations appear to be stable.

Maryland (no sport or commercial fisheries, classified as threatened, status extirpated) — Paddlefish are presently considered extirpated from the state (personal communication, 1983, Nick Carter, Fishery Biologist, Maryland Department of Natural Resources, Annapolis, Maryland).

Minnesota (no sport or commercial fisheries, classified as protected, status stable) — Paddlefish occur in the Mississippi, St. Croix and Minnesota rivers (personal communication, 1983, Bruce Hawkinson, Planning and Program Supervisor, Minnesota Department of Natural Resources, St. Paul, Minnesota). Historically, they occurred as far west as Mankato and in the St. Croix River up to Taylors Falls, but are no longer common in much of this range and have practically disappeared from many areas (Eddy and Underhill[144]). Paddlefish have reappeared in parts of their original range within Minnesota, and they are caught occasionally by commercial fishermen in Lake St. Croix and in the Mississippi River below Lake Pepin (Eddy and Underhill[144]). Around 1900 they were very abundant in Lake Pepin (Eddy and Underhill[144]).

Mississippi (sport and commercial fisheries, classified as commercial, status stable-increase) — Paddlefish are believed to occur in all watersheds in Mississippi. Present range is somewhat similar to the historical distribution (personal communication, 1983, C. A. Schultz, Fisheries Biologist, Mississippi Department of Wildlife Conservation, Aberdeen, Mississippi). Over 80% of the commercial harvest is from the Yazoo River watershed and the Mississippi River. Slightly less than 20% comes from the lower Pearl and lower Pascagoula rivers and Pickwick Lake, and not over 5% from the remainder of the state. According to commercial fishermen and law enforcement personnel, the paddlefish population appears to be increasing in the Yazoo River watershed from the low densities experienced after 1952. Similar increases have not been reported from other watersheds within Mississippi. Limited sport fisheries occur below flood control dams at Arkabutla, Sardis, Enid and Grenada on the Yazoo River.

Missouri (sport and commercial fisheries, classified as game, status stable) — The Mississippi, Missouri and Osage rivers presently support substantial populations of paddlefish. They also occur in lesser numbers in the Lamine, North Grand, St. Francis, Chariton and Gasconade rivers. Paddlefish have been found in most large rivers in the state, and their distribution has not changed except for a recent introduction into Table Rock Lake (Graham[218]). The largest paddlefish population and sport fishery of greatest significance is in the Lake of the Ozarks-Osage River (Purkett[409]). Recently in Table Rock Lake, another sport fishery has developed as a result of stocking hatchery-produced fingerlings (Graham[218]). Commercial fishing is allowed only in the Mississippi, Missouri and lower St. Francis rivers, and current harvests are much less than those prior to 1900. Populations are stable at present (personal communication, 1983, L. Kim Graham, Fisheries Research Biologist, Missouri Department of Conservation, Columbia, Missouri).

Montana (sport fishery only, classified as sport but designated as a species of special concern, status stable) — Paddlefish are found in Fort Peck Reservoir and the Missouri River above the lake, as far upstream as the Marias River, and in the Yellowstone River and Missouri River below Fort Peck Dam. The historical and present-day distributions are similar; however, those populations affected by Fort Peck and Garrison dams have been enhanced (personal communication, 1983, Al Elser, Chief, Fisheries Management, Montana Department of Fish, Wildlife and Parks, Kalispell, Montana). Present populations appear to be stable. Important sport fisheries occur at Intake Diversion Dam on the Yellowstone River, on the Missouri River above Fort Peck Reservoir, and at the Fort Peck Dredge Cuts below Fort Peck Dam.

Nebraska (sport fishery only, classified as sport, status stable) — The only significant population occurs in the Missouri River, but remnant populations occur in sandpits along the lower reaches of the Platte and Elkhorn rivers. With the exception of western reaches of the Platte River, distribution of paddlefish has not changed markedly in Nebraska, and present day status is stable (personal communication, 1983, Bud Stewart, Fisheries and Wildlife Biologist, Nebraska Game and Parks Commission, Norfolk, Nebraska). The sport fishery harvest below Gavins Point Dam appears to be increasing.

New York (no sport or commercial fisheries, classified as extirpated, status extirpated) — Newspaper accounts of a 123-lb. paddlefish from Lake Chautauqua represent the only recorded observation of this species in New York. They are considered extirpated from the state (personal communication, 1983, Bruce Shupp, Chief, Bureau of Fisheries, and George Seeley, District Supervisor of Fish Culture, Department of Environmental Conservation, Albany, New York).

North Carolina (no sport or commercial fisheries, not classified, status extirpated except for an unconfirmed report in 1983) — Paddlefish inhabited the French Broad River near Asheville during the late 1800s (Evermann[162]); however, personnel from the Wildlife Resources Commission have never observed paddlefish in North Carolina waters (personal communication, 1983, Red Harris, Program Manager-Inland Fisheries, North Carolina Wildlife Resources Commission, Raleigh, North Carolina). There was a report that a fisherman caught one during the week of Oct. 10, 1983, but this catch was not verified.

North Dakota (sport fishery only, classified as commercial, status stable-increasing) — Paddlefish are restricted to the Yellowstone and Missouri river systems in North Dakota, their historic distribution. The population of paddlefish above Garrison Dam is stable or may in fact be increasing, while the population downstream from Garrison Dam to the South Dakota border may have decreased (personal communication, 1983, Emil Berard, District Fisheries Manager, State Game and Fish Department, Riverdale, North Dakota). Commercial fishing is not permitted, but fish killed incidental to netting operations in Lake Sakakawea and Lake Oahe can be kept. Sport fishing is limited to that area on the Missouri River lying west of U.S. Highway 85 to the Montana border and that portion of the Yellowstone River in North Dakota.

Ohio (no sport or commercial fisheries, classified as endangered, status declining) — Paddlefish inhabited Ohio waters before 1780 (Trautman[512]) and were observed in the upper Ohio River and its larger tributaries prior to 1900. They were considered abundant in the Ohio River prior to 1915 but the population later decreased markedly (Trautman[512]). Confirmed reports of paddlefish taken from Lake Erie indicate a small population existed in Lake Erie prior to 1903. Paddlefish are not abundant in Ohio, although a few are observed regularly.

Oklahoma (sport and commercial fisheries, classified as commercial, status unknown) — The range of paddlefish in Oklahoma, which has changed little since the early 1900s, includes the Arkansas River, the Salt Fork of the Arkansas River to Great Salt Plains Reservoir, the Grand or Neosho River, the Verdigris River, the Canadian River to Eufaula Dam, the Cimarron Arm of Keystone Reservoir and the Red River to Texoma Reservoir (personal communication, 1983, David Combs, Fisheries Biologist, Oklahoma Department of Wildlife Conservation, Porter, Oklahoma). The overall status of paddlefish populations in Oklahoma is unknown; however, the Keystone Reservoir population has increased.

Pennsylvania (no sport or commercial fisheries, classified as extirpated, status extirpated) — Paddlefish were known to occur in Lake Erie and in the Allegheny and Clarion rivers, but they are now considered extirpated from the state (personal communication, 1983, Clark Shiffer, Endangered Species Coordinator, Pennsylvania Fish Commission, Bellefonte, Pennsylvania).

South Dakota (sport fishery only, classified as sport, status declining) — Paddlefish inhabit the Missouri River and four of its main-stem reservoirs. Populations developed in the tailwaters of Lake Oahe, Lake Sharpe and Lake Francis Case and then declined. The historical range has not changed but overall status of the population is one of decline (personal communication, 1983, Dennis Unkenholz, Missouri River Coordinator, South Dakota Department of Game, Fish, and Parks, Pierre, South Dakota). The largest sustained sport fishery now occurs below Gavins Point Dam on the Missouri River.

Tennessee (sport and commercial fisheries, classified as commercial, status stable) — The Tennessee and Cumberland river drainages support paddlefish populations and their status is stable (personal communication, 1983, James Earl Hayes, Fisheries Biologist, Tennessee Wildlife Resources Agency, Murfeesboro, Tennessee). Paddlefish no longer occur in Reelfoot Lake, and none have been taken from Dale Hollow and J.P. Priest reservoirs since impoundment. The largest commercial harvest in the Tennessee River system is from Kentucky Lake on the Tennessee River, and this fishery may be increasing. Sport fisheries exist below Center Hill Reservoir on Caney Fork of the Cumberland River and in the tailwaters of Great Falls Reservoir (above Center Hill Reservoir) near Rock Island. There is no concentration at the Tellico Dam site on the Little Tennessee River.

Texas (no sport or commercial fisheries, classified as endangered, status declining) — According to Burr[86], paddlefish are found in the Red, Sabine, Neche and San Jacinto rivers and Cypress Creek (Caddo Lake). Paddlefish are either severely depleted or extinct in drainages of Galveston Bay and Sabine Lake. Commercial fishermen have reported taking paddlefish from the Sulphur River system (Bonn and Inman[56]), and fish were also reported from Lake Anahuac (personal communication, 1983, Alan Wenger, Regional Director-Inland Fisheries, Texas Parks and Wildlife Department, Austin, Texas). Paddlefish are very rare or extinct in the Trinity River. The report of paddlefish in the San Jacinto River (Burr[86]) is questionable (personal communication, 1983, Alan Wenger, Regional Director-Inland Fisheries, Texas Parks and Wildlife Department,

Austin, Texas). The overall perspective indicates that paddlefish are almost absent from Texas and that population levels have certainly declined.

Virginia (sport and commercial fisheries, classified as non-game, status stable) — Paddlefish occur in the Clinch River flowing into Norris Reservoir (personal communication, 1983, Jack Hoffman, Chief, Fisheries Division, Virginia Commission of Game and Inland Fisheries, Richmond, Virginia). They are not abundant and the historic range has remained constant.

West Virginia (sport fishery only, classified as sport but listed as threatened, status declining) — Paddlefish are occasionally observed in the Ohio River (personal communication, 1983, Fred Lekie, Fishery Biologist, West Virginia Department of Natural Resources, Charleston, West Virginia). Paddlefish were considered abundant prior to 1900 but have since declined (personal communication, 1983, Bernie Dowler, Administrator-Warmwater Fisheries, West Virginia Department of Natural Resources, Charleston, West Virginia) although they still occupy approximately the same range.

Wisconsin (no sport or commercial fisheries, placed on watch list, status stable) — Paddlefish are present in the Mississippi and Wisconsin rivers, but not above Prairie du Sac Dam on the Wisconsin River. The distribution of paddlefish in Wisconsin has not changed appreciably and the current population appears to be stable (personal communication, 1983, Jim Holzer, Fisheries Biologist, Wisconsin Department of Natural Resources, LaCrosse, Wisconsin).

Summary

Paddlefish populations and fisheries have changed markedly since the turn of the century. Significant reductions of principal paddlefish stocks have occurred in the Mississippi, Missouri, Ohio and Red rivers, while notable increases in stock size have occurred in the Tennessee, Cumberland and Arkansas rivers. The peripheral range of paddlefish has diminished slightly; however, paddlefish still occur in the majority of their original range, and in some instances may be reinvading areas previously occupied. They still occur in 22 states.

Commercial harvest has decreased by approximately 60% since the early 1900s and individual fisheries have declined by as much as 95%. Commercial fisheries in the Tennessee, Cumberland and Arkansas rivers now account for almost 70% of the reported catch in the United States. Based upon commercial catch statistics, the largest paddlefish fisheries presently exist in the Tennessee, Mississippi, Cumberland and Arkansas rivers.

Paddlefish have been extirpated from four states and all Canadian waters. Four states (Minnesota, Ohio, Texas and Wisconsin) now protect paddlefish from all exploitation.

Acknowledgments

Special recognition must be given to Douglas Carlson and Pat Bonislawsky for their early treatment of this topic. The present status of paddlefish populations and fisheries could not have been evaluated without the assistance of numerous colleagues and resource agencies throughout the United States.

Appreciation is extended to Jim Mayhew, Marion Conover and Larry Mitzner for reviewing the manuscript and my co-workers in the Paddlefish Work Group for early consultation. The constant encouragement and willing counsel of L. Kim Graham and Thomas R. Russell is also appreciated.

Appendix A: A questionnaire to determine paddlefish occurrence, abundance, present status, classification, and impacts of habitat changes on paddlefish sport fisheries.

1. Describe present paddlefish range in your State. Please indicate where sport, commercial or no harvest occurs (a State map indicating which river systems contain paddlefish would be helpful).
2. Define historical paddlefish distribution in your State if different from your present distribution.
3. What is the present status of paddlefish in your State? Is the status stable, increasing or decreasing on all or part of its range? Comments and data concerning changes in abundance vs time are especially useful. Three time periods could be considered: 1) early 1900s, 2) mid-1900s, and 3) present time.
4. Has abundance changed since construction of reservoirs? Have the populations increased or decreased? By what extent?
5. What is the present classification of paddlefish (sport or commercial) and what are the legal sport harvest methods? (A copy of your State fishing regulations should be sufficient; add comments if pertinent.) Describe, if any, changes in sport fishing techniques over the years.
6. What were approximate 1982 harvest estimates of each paddlefish sport fishery in your State? (Include number and mean weight.)
7. Has overharvest by sport fishing occurred on all or any portion of the paddlefish range in your State? If so, cite literature documenting overharvest or describe here.
8. Describe any paddlefish population changes due to environmental impacts (cite literature if appropriate).
9. Describe effects of environmental conditions such as water temperature, flow rates and volumes, change in flows, etc., on harvest success.

Appendix B: State personnel contacted by telephone to complete the paddlefish questionnaire.

Alabama — William C. Reeves, Department of Conservation and Natural Resources, Montgomery, Alabama.

Arkansas — Mike Freeze, Arkansas Game and Fish Commission, Lonoke, Arkansas.

Illinois — Dan Sallee, Illinois Department of Conservation, Aledo, Illinois. Jerry Rasmussen, U.S. Fish and Wildlife Service, Rock Island, Illinois.

Indiana — Joe Janisch, Indiana Department of Natural Resources, Indianapolis, Indiana.

Iowa — Tom Gengerke, Iowa Conservation Commission, Spirit Lake, Iowa.

Kansas — Bruce Taggart, Kansas Fish and Game Commission, Chanute, Kansas.

Kentucky — Ted Crowell, Kentucky Department of Fish and Wildlife Resources, Frankfort, Kentucky.

Louisiana — Richard Brassette, Department of Wildlife and Fisheries, Baton Rouge, Louisiana.

Maryland — Nick Carter, Maryland Department of Natural Resources, Annapolis, Maryland.

Minnesota — Bruce Hawkinson,

Minnesota Department of Natural Resources, St. Paul, Minnesota.

Mississippi — C.A. Schultz, Mississippi Department of Wildlife Conservation, Aberdeen, Mississippi.

Missouri — Joe Dillard, L. Kim Graham and Thomas R. Russell, Missouri Department of Conservation, Columbia, Missouri.

Montana — Al Elser, Montana Department of Fish, Wildlife and Parks, Kalispell, Montana.

Nebraska — Bud Stewart, Nebraska Game and Parks Commission, Norfolk, Nebraska.

New York — Bruce Shupp and George Seeley, Department of Environmental Conservation, Albany, New York.

North Carolina — Red Harris, North Carolina Wildlife Resources Commission, Raleigh, North Carolina.

North Dakota — Emil Berard, State Game and Fish Department, Riverdale, North Dakota. Jim Ragen, State Game and Fish Department, Bismarck, North Dakota.

Ohio — Information taken from Trautman[512].

Oklahoma — Dave Combs, Oklahoma Department of Wildlife Conservation, Porter, Oklahoma.

Pennsylvania — Clark Shiffer, Pennsylvania Fish Commission, Bellefonte, Pennsylvania.

South Dakota — Dennis Unkenholz, South Dakota Department of Game, Fish, and Parks, Yankton, South Dakota.

Tennessee — James Earl Hayes, Tennessee Wildlife Resources Agency, Murfreesboro, Tennessee, and Ron Pasch, Tennessee Valley Authority, Division of Water Resources, Knoxville, Tennessee.

Texas — Bob Kemp and Alan Wenger, Texas Parks and Wildlife Department, Austin, Texas.

Virginia — Jack Hoffman, Virginia Commission of Game and Inland Fisheries, Richmond, Virginia.

West Virginia — Bernie Dowler and Fred Lekie, West Virginia Department of Natural Resources, Charleston, West Virginia.

Wisconsin — Jim Holzer, Wisconsin Department of Natural Resources, LaCrosse, Wisconsin, and Vern Hacker, Wisconsin Department of Natural Resources, Oshkosh, Wisconsin.

Threats to Paddlefish Habitat

Rollin D. Sparrowe

U.S. Fish and Wildlife Service
U.S. Department of the Interior
Washington, D.C. 20240

Abstract

Paddlefish (*Polyodon spathula*) are big-river inhabitants, and require certain flows, temperatures and substrate conditions for growth and reproduction. As man modified the big-river systems of North America, he adversely affected this species.

Water resource projects such as dredging, flow modification, and dam and reservoir construction have altered most of the original paddlefish habitat in the United States. Although paddlefish have benefited from some of these projects, in general the associated structural changes in the rivers have resulted in dramatic losses of habitat and significant decreases in paddlefish populations. In addition, the operation of these projects may be one of the biggest threats to paddlefish in the future.

Many examples exist of the influence of water projects on paddlefish. For instance, dams on the Tennessee, Osage, Yellowstone and Missouri rivers have had major effects on paddlefish populations. Fisheries flourished at first, then declined drastically because of a lack of recruitment.

Historical records clearly reveal that many problems paddlefish face today were caused by incremental development of water resources throughout their range. No one planned to systematically eliminate their spawning sites or disrupt their spawning behavior and migration, but these were the biological effects of gradual, incremental modification of big rivers and their watersheds. Planning agencies did what seemed right at the time; unfortunately, the cumulative effects on paddlefish have only recently become evident.

The future of paddlefish is only one of many issues that must be considered in natural resource planning for big rivers. However, the condition of paddlefish populations is a good indicator of the overall condition of our big-river resources. The best hope for paddlefish rests in our ability to stick to a revised approach to water development, while maintaining current uses of big rivers. Otherwise paddlefish will suffer the same fate as the Pacific salmon populations in the Columbia River system, which soon may have to be maintained with hatchery-raised recruits.

Although we have the technical expertise to raise paddlefish in hatcheries, we hope it will never be required to sustain populations. This is a view many of us share, and one that makes our efforts to save the remaining natural paddlefish habitat all the more urgent.

The native range of paddlefish (*Polyodon spathula*) includes some of the largest and most economically important big-river systems in North America. These river systems (the Missouri, Ohio and Mississippi rivers and their tributaries) supply much of the United States with water resources for commercial, industrial and municipal uses.

Over the past 60 years, man has

drastically modified these river systems by dam and reservoir construction, dredging, flow modification and pollution. Increasing evidence suggests that these changes have had serious consequences for paddlefish.

Paddlefish require a fairly narrow range of river flow, temperature and substrate conditions for reproduction and other stages of their life cycle (Russell[446]). Structural changes in big-river systems have altered these natural conditions in many areas, and it is likely that most of the original habitat has been adversely affected. The history of these structural changes is the focus of this paper.

Although an important sport fish in some localities, paddlefish have never had the widespread popularity of other river species. As a result, they received little attention from researchers. Only within the past decade or so has basic biological information been gathered on these fish. Interest is increasing, however, and it appears that the paddlefish has finally captured the attention of both biologists and the public because of its uniqueness, its potential economic and recreational value, and its dramatic decline in abundance in many areas.

Unfortunately, this change in public awareness occurred long after major commercial and industrial uses of the rivers had been established. Any program protecting paddlefish habitat must take into account these established uses. Managers must also be alert for new commercial and industrial developments that might affect paddlefish in the future. As with most modern resource management issues, protecting paddlefish habitat will require both an assessment of public and official awareness of the resource, and an evaluation of ways to sustain that resource in the face of changing priorities most often set by non-resource management interests.

Missourians have witnessed drastic modifications of the habitat of perhaps the largest population of paddlefish in North America. Paddlefish in the Osage River, Missouri, have economic and recreational value as well as cultural importance. Fishermen and their families traditionally gather each year to harvest these fish (Purkett[409]; Russell[450]; Graham[217]). Unfortunately, the Osage River paddlefish population was seriously affected by the loss of spawning habitat upon completion of Harry S. Truman Dam in 1977 (Russell[450]; Graham[217]).

The effects of Truman Dam on paddlefish became a highly sensitive issue in Missouri. This level of concern on the part of both biologists and the public has been reached in few other states, so it is appropriate that we discuss threats to paddlefish habitat at this symposium here in Missouri.

Although this paper refers often to big rivers as "paddlefish habitat," we should remember that the paddlefish is only one of many species that inhabit and depend upon big-river systems, and only one of many resources that are affected by the human use of these rivers.

Projects Which Affect Paddlefish Habitat

Many people understandably associate paddlefish with reservoirs rather than rivers, because significant fisheries have developed in many reservoirs and tailwaters of reservoirs throughout the fish's range. Although paddlefish have prospered in some reservoirs, in general the structural modification of big rivers — primarily through construction of reservoirs and channelization for both flood control and navigation — has resulted in dramatic losses of paddlefish habitat. Such water development projects cause major physical changes in the rivers, and their operation may be one of the biggest threats to paddlefish in the future.

Dams on the Tennessee, Osage, Yellowstone, and Missouri rivers have had major effects on paddlefish populations. Following dam closure, paddlefish on spawning runs congregated in the tailwaters below the dams. Fishing success in these areas was temporarily improved over that found in natural river systems. Several fisheries boomed for a number of years, and then declined drastically because of lack of recruitment (Pasch and Alexander[390]; Unkenholz[528]). The challenge to fisheries managers is to determine how to maintain paddlefish populations and fisheries in such situations.

Reservoir construction has caused specific problems for paddlefish in some cases, and produced benefits in others. Loss of spawning habitat, blocked migration routes, altered flow regimes and hydrology, and direct kills through operation of hydropower facilities are among the problems confronting paddlefish at a dam site. Benefits include temporarily increased fishing opportunities and greatly improved and expanded feeding areas within reservoirs (Russell[446]; Graham[218]).

Gross structural changes made to benefit navigation have had detrimental effects on habitat in all major river systems in the original paddlefish range. These changes include the construction and maintenance of deeper channels, alteration of bank stability, channelization, and modification of flow regimes at different times of the year to maintain navigation. The biological impacts of these water development projects on paddlefish fisheries are addressed by Unkenholz[528].

Historical Development of Habitat Changes

One of the earliest major plans for water projects that would affect paddlefish habitat was published in 1937. This document, "Drainage Basin Problems and Programs," was developed by the National Resources Committee for presentation to President Roosevelt. It divided the United States into 17 major river basin areas, including the Ohio, the Upper Mississippi, and the Missouri, and assembled water project recommendations from several resource planning efforts of the early 1930s.

The Pick-Sloan Plan for the Missouri Basin was proposed in the early 1940s to control floods on the Missouri and Mississippi rivers (U.S. Congress, Senate 1944). This large-scale planning effort accelerated the earlier proposed plans. Reservoirs built in Missouri and on major rivers that converge near St. Louis were products of Pick-Sloan and earlier planning efforts, and of local plans such as the Missouri Water Plan (Missouri State Planning Board 1938).

The project justifications in these and similar early documents certainly did not consider the needs of paddlefish or the effects on their habitats. Unfortunately, these early plans clearly led to the modifications of habitat that paddlefish managers are confronted with today and will have to cope with in the future. Maps and planning statements in the 1937 document and in specific plans for developing the Missouri River (U.S. Congress, House 1934) show proposed dam, channel, and irrigation projects that are strikingly similar to those that were constructed 30 and 40 years later.

Paddlefish weren't the only species overlooked by water project planners in the 1930s and 1940s. A close examination of these early plans reveals only a rudimentary appreciation of fish and wildlife resources in general.

To be fair, fish and wildlife agencies were just getting started at this time, and were no more accustomed to long-range planning than were development interests. While we might criticize early planners for not accurately predicting the current impacts of water projects upon the fish and wildlife resource, we should remember that long-term planning is always difficult. In the long interval between planning and construction of water projects, many things changed, including public attitudes, government priorities, population centers, and the amount of biological information available on paddlefish and other river species. Ideally, long-term plans should be subject to periodic review and modification as new information becomes available, but this does not always happen.

Historical records reveal the problems caused by incremental development of water resources throughout the range of paddlefish. No one planned to systematically eliminate paddlefish spawning sites, disrupt paddlefish spawning behavior and migration, or pollute paddlefish habitat with sediment and municipal waste. These effects all came about gradually as big rivers and their watersheds were developed. Planning agencies simply did what seemed to be right at the time to provide for the country's varied water needs, both current and anticipated. The overall threat to paddlefish became evident only recently, after the cumulative effects of these projects began to adversely affect paddlefish populations.

The chronology of water development, and its impacts on paddlefish and many other fish and wildlife species throughout the Mississippi River drainage, is a classic example of resource management evolution in America. First came Pick-Sloan, largely for flood control and later for power generation on the Missouri River; then locks and dams on the Mississippi River for flood control and navigation; then similar modifications on the Ohio and other major river systems. Tremendous momentum gathered to "develop" water resources. Social, economic and natural resource conditions changed at each stage in development of these projects, as did justifications. Early justifications were flood control and hydroelectric power; later recreation, especially fishing, became an important reason for more river modifications. Now we are in a period of concern for the biological resources of big rivers, and are confronted with the damage done by previous development.

Case Histories

It is difficult to say whether channel and flow modification of large rivers or construction of reservoirs has had the more profound effect on paddlefish. Both have eliminated important spawning habitat, but both have resulted in some benefits to paddlefish populations. Case histories will illustrate these points.

Case History 1: Big River Channel Modification

Through the Missouri River Bank Stabilization and Navigation Project, the Missouri River has been channelized from its mouth at St. Louis to near Ponca, Nebraska (approximately 750 river miles). Only a 55-mi. stretch of unchannelized river remains below the first main-stem reservoir at Gavins Point Dam, South Dakota. Channelization has reduced the surface area of the river by about one-half, yet the velocity has nearly doubled. Previous interspersions of riffles, pools and backwaters with islands and sandbars, and the ever-shifting channel flows, have been almost totally eliminated. The mechanics and impacts of these changes are described by Funk and Robinson (1974).

Six main-stem reservoirs on the Missouri River (Fig. 1) provide water storage, allow flow manipulation downstream, produce hydropower, and contribute in various ways to paddlefish and other fisheries. In their 1982-83 Operating Plan for the Missouri River main-stem reservoirs, the U.S. Army Corps of Engineers states that "the construction of the main-stem reservoirs has been one of the most important contributions to sport fishing in the Missouri Basin in the century." The narrative also describes significant fisheries for northern pike (*Esox lucius*), walleye (*Stizostedion vitreum*), sauger (*Stizostedion canadense*) and paddlefish.

There have been trade-offs, however, in receiving these benefits. While good paddlefish populations have developed in some reservoirs due to enhanced zooplankton production, many of those populations have had problems maintaining themselves because of lack of spawning habitat and recruitment (Unkenholz[528]). Further, flow management for flood control and power production could threaten zooplankton production in the reservoir. Finally, whatever "natural" habitats are left in the river are not likely to benefit from flow regime modification.

Like the Missouri River, the Mississippi River has been dramatically

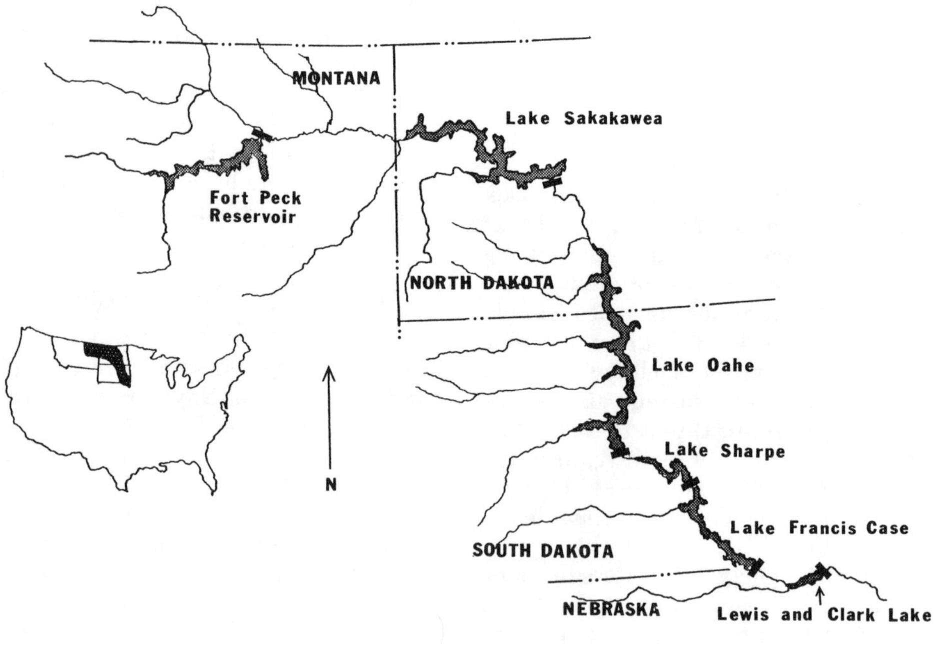

Figure 1. *The six main-stem reservoirs on the Missouri River.*

altered by water projects. Almost total modification of the river from Minneapolis to St. Louis through a series of 27 locks and dams has produced serious problems for paddlefish. Early information available on paddlefish populations in the Mississippi River indicates that there has been an overall decline in paddlefish numbers since lock-and-dam construction (Carlson and Bonislawsky[95]; Gengerke[196]).

Locks and dams facilitate navigation on the river by stabilizing pool levels. Pool stabilization, along with dredging and flow manipulation, has decreased spawning habitat for paddlefish and reduced the habitat diversity of the entire Mississippi River. The impacts of these modifications are more serious in the southern portions of the Mississippi River Basin. Similar changes have occurred on the Ohio River and on many other large-river tributaries of the Mississippi River.

Case History 2: Dam Construction on the Osage River

The positive and negative effects of reservoir construction on paddlefish are well-demonstrated by what has occurred in the Osage River Basin, mainly in Missouri (U.S. Congress, House 1950).

Lake of the Ozarks, formed by construction of Bagnell Dam in 1931, was the first reservoir on the Osage River. A paddlefish population developed from fish trapped above the dam. Factors apparently responsible for this development were favorable feeding conditions in the reservoir and spawning habitat in the Osage River above the reservoir. Suitable spawning conditions may be the most important factor determining whether a significant paddlefish population will be maintained following impoundment. Paddlefish populations that developed in reservoirs lacking spawning habitat frequently declined rapidly (Unkenholz[528]).

Like other paddlefish populations in reservoirs, the Lake of the Ozarks population first received attention when fishermen began catching large numbers of paddlefish. In this case, however, the fish were being caught above a reservoir, rather than in the tailwater. This resulted in several decades of a unique kind of recreation, one that brought families from surrounding areas to the Osage River to harvest these large fish during spring water rises. A cultural tradition of catching and eating paddlefish soon developed.

The next two large reservoirs, Pomme de Terre and Stockton, were constructed in 1961 and 1969, respectively, on tributaries of the Osage River in Missouri without apparent effect on paddlefish. However, when Harry S. Truman Dam was constructed and the reservoir finally filled in 1977, major destruction of paddlefish spawning habitat occurred. This dam inundated a 60-mi. length of river containing all known paddlefish spawning habitat in the Osage River Basin (Russell[450]), and blocked spawning runs out of Lake of the Ozarks.

Truman Dam has recently caused other problems for paddlefish. Paddlefish and other species have been killed by supersaturation due to high spillway releases, and by physical damage caused by pump-back testing and spillway modification.

Supersaturation problems resulted in high fish mortalities in 1978 and 1979 (Crunkilton et al. 1980). Some paddlefish were killed, but the main impact was on many other species of fish. In a similar incident in the spring of 1982, the first test of one of the five turbines for the hydropower pump-back unit killed large numbers of fish, and created a political furor. During this 3-hour test on April 13 and 14, 1982, at least 15 species of fish, including paddlefish, were killed by being drawn into the turbines during the pump-back operation. An estimated 2,000 lb. of dead and mutilated fish floated to the surface immediately after the pumps were started (Richards[421]). While it was difficult to estimate the number of fish lost, the fish kill was dramatic.

Paddlefish were also killed in the springs of 1980 and 1983 by a suspected "shear plane" effect, apparently caused by a combination of high-water releases and spillway design. When water temperatures approach 55 F and flow releases over the floodgates exceed 20,000 cubic feet per second (cfs), paddlefish concentrate in large numbers in the tailwaters. When these conditions occur, several hundred paddlefish are badly injured or killed within a few days (personal communication, 1983, L. Kim Graham, Fisheries Research Biologist, Missouri Department of Conservation, Columbia, Missouri).

Such experiences in Missouri and elsewhere have shown that inflexible engineering and operation of dams at hydropower units can cause major problems for paddlefish and other species. The pump-back test in April 1982 was conducted despite repeated public and official expressions of concern about such testing in the spring, when fish on spawning runs congregate below the dam.

In addition to killing fish directly, the 1982 test had a disastrous effect on walleye. In March a large volume of water was discharged into the tailwaters to lower the reservoir pool for pump-back testing. This attracted a spawning run of walleye into the tailwater area. These high water releases were followed by severely reduced flows preparatory to pump-back testing, which significantly lowered the water level in the tailwater area. Because of receding water levels, the walleye returned to the Lake of the Ozarks after a badly disrupted spawning effort (Richards[420]). Long-range impacts on this important developing fishery will certainly reduce one of the anticipated benefits from outflows from Truman Reservoir.

If there is one thing water resource planners are unable to promise, it is a consistent water regime. Since dams such as Truman are built primarily to hold back water in order to prevent floods or to produce hydropower, it should be no surprise that guarantees as to time, amount and duration of waterflows will be difficult to meet. All of the flow problems that have occurred to date at Truman Dam were forecast in documents made public in the early 1970s during controversy over completion of the reservoir (U.S. Army Corps of Engineers 1973). Most of these problems are documented in a case history of that reservoir (Sparrowe[479]).

Although Truman Dam continues to cause problems for paddlefish and other species, the Corps of Engineers has stopped the pump-back operation because of its damaging effect on the fishery. I hope this is a sign that the management inflexibility we have witnessed in the past is being replaced by a concern for long-range stability of fish populations in Truman Reservoir.

Releases from Stockton Lake, which flow directly into Truman Reservoir, have illustrated that there are continuing problems when water regimes are influenced by needs for flood control or power generation. Missouri reservoirs are not the only ones with such problems: the first floods below Shelbyville and Carlyle reservoirs in Illinois demonstrated the difficulty of planning and managing flows from large reservoirs. The inability of flood control projects to provide stable and predictable flows remains one of the most difficult of many problems posed for managers of paddlefish and other fish species below reservoirs in the Midwest.

Other Threats to Paddlefish Habitat

Along much of the Missouri, Ohio and Mississippi rivers and their many tributaries, agricultural development has contributed to drastic changes in the rivers.

This country has gone from acute awareness of the erosion problem in the 1930s and 1940s to mismanagement of soil resources and reduction of erosion control efforts in the 1960s and 1970s. Soil movement and deposition are degrading our major rivers, and fertilizers and pesticides are contaminating them. Further, proposals now exist to pipe water from rivers or mainstem reservoirs across state boundaries for large-scale irrigation or other uses. Downstream residents are concerned about the impacts these water diversions may have on fisheries, recreation and water availability.

The water quality of many rivers has been drastically changed by watershed development. Industrial pollution is severe along the Ohio River and other tributaries to the Mississippi River, and problems with municipal wastes occur at many sites. Of current concern are the funding crises in many cities, which may cause a reduction in efforts to control water pollution problems.

Alterations in habitat and reductions in water quality due to siltation and pollution pose problems to paddlefish populations in many states. Larger economic and social needs than fishing or recreation caused these problems, and it will take regional and national efforts to overcome them.

Changes in National Water Policy and Direction

The passage of the National Environmental Policy Act in 1969, the beginning of the Environmental Protection Agency in 1970, and the initiation of an array of water resource needs assessments and policies during the 1970s, had major positive impacts on national water policy. During this period, administrators tried to coordinate water development programs, and concerned citizens took advantage of NEPA's public expression requirements. Everyone began to realize how many natural resource issues had been affected by incremental development of our river systems. The 1973 Report of the National Water Commission, entitled "Water Policies for the Future," outlined the socio-economic conflicts over water developments that had characterized previous decades.

Also during the 1970s, the government began expecting more equitable cost-benefit analyses for water development projects, requiring more direct economic participation by local beneficiaries, and adopting a "user pays" attitude toward navigation projects. In other words, it began to take a more reasoned approach by balancing water development needs with other resource benefits.

Many battles remain to be fought over justification of new projects. In the meantime, threats to paddlefish and other species may be lessened by improving the operation, and possibly the structure, of existing large-river modification projects.

Since 1976, no new water resource projects have been approved by Congress. The Carter Administration in the late 1970s adopted a "hit list" of projects considered overly expensive or unnecessary. The current administration has received wide support for encouraging local payment of construction costs and for supporting a user-pays approach to maintenance and operation.

Although these moves seem to be in the right direction, water issues are still controversial. In fact, omnibus bills including new water developments have recently been introduced in Congress. These bills propose nearly 100 new projects, and contain a policy which specifies that states or other beneficiaries must pay at least 35% of the costs of multiple-purpose projects, and up to 100% for projects built solely for local benefit.

Reactions to these bills have varied. Although paddlefish and other fishery resources will obviously benefit from careful water policy planning, such plans often change or fail. Even the 1937 planning document for water resources (National Resources Committee 1937) ascribed to basic principles such as "user pays." We seem to have trouble sticking to our guns on such philosophies.

Within an hour's drive of St. Louis is the site of another episode in water resource decision-making that could affect the future of paddlefish habitat. The Meramec Dam project was finally deauthorized in 1981, after a long public and political battle, much of it fought in St. Louis. There are dozens of similar projects proposed for Missouri, and hundreds more for other states — projects which have been on the books since those early planning studies in the 1930s. One way of maintaining habitat for paddlefish and many other species would be deauthorization of these projects, as occurred with the Meramec project. This would allow new, long-range planning for natural resources within those watersheds. Those who participated in the campaign to deauthorize the Meramec Dam, however, know how much personal and political energy such an effort takes.

Prognosis

The paddlefish is only one of many large-river resources which must be considered in natural resource planning. However, the condition of the paddlefish population can be regarded as an indicator of the overall condition of our water resources. The hope for the future of paddlefish lies partly in our ability to stick to a revised approach to water development and to maintain current uses of our large rivers. The challenge for fishery managers will be to maintain the paddlefish population within the modified habitats of today's big rivers.

I cannot help but be concerned that paddlefish will suffer the same fate as the various species of Pacific salmon in the Columbia River and its tributaries. Unlike paddlefish, Pacific salmon had the benefit of early public interest and commercial value. Nevertheless, incremental development of the water resources of the Columbia River and its tributaries have drastically reduced salmon populations thought limitless only a few decades ago. In many areas of the Northwest, salmon may soon have to be hatchery-raised to maintain runs for both sport and commercial take. Fall-run strains of chinook salmon (*Oncorhyncus tshawytscha*) have even been informally suggested for listing under the Endangered Species Act.

The parallel with paddlefish and their habitat is fairly easy to draw. An article in Audubon magazine (Gilbert[204]) on paddlefish gave public expression to some of the concerns voiced here today. The article praised resource managers for overcoming technical problems associated with raising paddlefish in hatcheries, but expressed sorrow at the necessity of doing it at all. It's a feeling we all share, and one that makes our efforts to save the remaining natural paddlefish habitat all the more urgent.

References

CRUNKILTON, R. L., J. M. CZARNEZKI, AND L. TRIAL. 1980. Severe gas bubble disease in a warmwater fishery in the midwestern United States. Transactions of the American Fisheries Society 109(6):725-733.

FUNK, J. L. AND J. W. ROBINSON. 1974. Changes in the channel of the lower Missouri River and effects on fish and wildlife. Missouri Department of Conservation, Aquatic Series No. 11, Jefferson City, Missouri, USA. 52 pages.

MISSOURI STATE PLANNING BOARD, 1938. A state plan for Missouri, water plan 1938. Jefferson City, Missouri, USA. 380 pages.

NATIONAL RESOURCES COMMITTEE. 1937. Drainage basin problems and progress. Report by the National Resources Committee to the President. U.S. Printing Office, Washington, D.C., USA. 546 pages.

U.S. ARMY CORPS OF ENGINEERS. 1973. Harry S. Truman Dam and Reservoir environmental statement. Department of the Army, Corps of Engineers, Kansas City District, Kansas City, Missouri, USA. Including Appendixes A-G.

U.S. CONGRESS, HOUSE. 1934. Missouri River. 73rd Congress, 2nd Session, House Document 238. 1,245 pages.

U.S. CONGRESS, HOUSE. 1950. Osage River and tributaries, Missouri and Kansas. 81st Congress, 2nd Session, House Document 549. 33 pages.

U.S. CONGRESS, SENATE. 1944. Missouri River Basin. 78th Congress, 2nd Session, Senate Document 191. 211 pages.

Effects of Commercial Fishing on Paddlefish Populations

Ronald W. Pasch and C. Michael Alexander
Tennessee Valley Authority
Office of Natural Resources and Economic Development
Knoxville, Tennessee 37902

Abstract

The commercial catch of paddlefish (*Polyodon spathula*) has fluctuated since the late 1800s, varying with demand for roe and smoked flesh, which are often substituted for sturgeon caviar and smoked sturgeon. Harvest of paddlefish peaked around 1900, after depletion of lake sturgeon (*Acipenser fulvescens*) and Atlantic sturgeon (*Acipenser oxyrhynchus*) populations. This peak was followed by a long period of declining harvest. During World War II, harvest increased temporarily due to reduced imports of sturgeon flesh, and then declined through the 1950s and 1960s. In the 1970s, the value of roe increased to over $20/lb., and paddlefish harvest again rose dramatically.

Certain behavioral characteristics — schooling tendencies, predictable spawning runs, and winter distribution — make paddlefish fairly easy to locate and capture, and have contributed to the depletion of many commercially exploited populations. Susceptibility of paddlefish to capture is of particular concern because their population turnover rate is low compared with most freshwater fishes, and they are unable to sustain high levels of exploitation.

Commercial exploitation has been a major factor affecting paddlefish populations in several southern reservoirs. Case histories provide substantial evidence that commercial harvest can rapidly deplete adult paddlefish stocks. On the other hand, paddlefish populations can withstand moderate fishing pressure. Reports indicate that annual exploitation in sport and mixed sport/commercial fisheries of 15-20% does not damage these fisheries.

Although no paddlefish population has been extirpated solely by excessive exploitation, recent increases in roe value provide an incentive for overharvest. Therefore, paddlefish stocks should be managed conservatively, with regular evaluation of the effects of controlled harvest on population age and size structure.

In the late 1800s, paddlefish (*Polyodon spathula*) had little economic importance, although commercial harvest was significant. Coker[104] noted approximately 1 million lb. of paddlefish were harvested in 1894, but the catch was valued at only $21,000. By 1899, harvest exceeded 2.4 million lb., the greatest commercial catch recorded for this species.

During this period, paddlefish were sought primarily for their flesh, although their roe added considerable value. In 1898 paddlefish roe brought commercial fishermen 35-40 cents/lb. in Louisville, Kentucky, and somewhat more in New York (Hussakof[259]). Stockard[489] reported smoked paddlefish was marketed as sturgeon, and paddlefish eggs were mixed with sturgeon eggs in caviar production.

During the first two decades of this century, paddlefish caviar rose in price to $1.50-$3.00/lb., equalling that of sturgeon (Alexander[10]; Coker[104]). Surprisingly, at this time the Pure Food and Drug Act required that all caviar, except sturgeon, be designated by the name of the fish of origin. The price increase was probably due in large part to decimation of lake sturgeon (*Acipenser fulvescens*) and Atlantic sturgeon (*Acipenser oxyrhynchus*) stocks, and to decreasing commercial harvests of paddlefish (approximately 1.4 million lb. in 1903). Several paddlefish populations were quickly depleted, and biologists expressed concern about survival of these stocks (Stockard[489]; Alexander[10]; Coker[104]). At the 1914 meeting of the American Fisheries Society, Dyche and Ward (Alexander[10]) warned that paddlefish, like sturgeon, might be locally extirpated by commercial fishermen harvesting them for roe.

Paddlefish harvest statistics were reported infrequently until 1955 (Carlson and Bonislawsky[95]); however, these records reveal that the commercial catch fell 30% between 1903 and 1950. Depletion of easily harvested populations in lakes, along with declining demand for paddlefish roe, probably accounted for an overall decrease in harvest during this period.

During World War II, however, reduced imports of sturgeon flesh stimulated harvest of paddlefish for several years. Over 700,000 lb. of paddlefish were harvested from Wheeler and Wilson reservoirs, Alabama, and sold in large city markets during 1941 and 1942 (Bryan[76]). The nationwide commercial catch statistics for paddlefish in 1942 were not reported, but may have approached levels recorded early in the century. Some paddlefish caviar was undoubtedly being produced, but probably did not command the prices it once did.

From 1965 to 1975, the average nationwide annual paddlefish harvest was over 500,000 lb. (Carlson and Bonislawsky[95]). During this period, paddlefish roe ranged from 25 cents to $2/lb., and apparently many commercial fishermen discarded it due to these low prices. By spring 1979, roe (salted and bulk packed) had increased in value to $12-$15/lb. However, in many areas roe was still discarded, because fishermen were not yet aware of the developing market.

This situation soon changed. When fishermen realized that large, gravid females were worth more than $200 each for roe alone, commercial pressure quickly increased in the Tennessee, Cumberland, and Alabama river systems. Over 750,000 lb. dressed weight of paddlefish were harvested from impoundments on the Cumberland and Tennessee rivers during 1980 (Sample[451]). This estimate is probably conservative, since neither fishermen nor dealers systematically reported catches or purchases, and poaching may have been significant.

Demand for paddlefish roe has spread quickly from the Tennessee Valley region and caused alarm within several state conservation agencies, including those managing paddlefish for sport harvest. If roe prices remain at present levels or increase, pressure on paddlefish populations will likely increase as well.

In this report, we will describe the impacts of commercial harvest on paddlefish stocks, and suggest strategies for managing paddlefish that will protect populations during periods of high commercial fishing pressure.

Potential for Stock Depletion

For a fish stock to support a stable fishery, losses from mortality must be balanced by gains from growth and recruitment. In many exploited paddlefish populations, this balance could not be achieved. Commercial exploitation resulted in noticeable reductions in catch, and some fisheries were abandoned because they became unprofitable.

Paddlefish populations are particularly vulnerable to overexploitation for two reasons: first, because their behavior makes them easy to catch with commercial fishing methods; and second, because their relatively low recruitment and growth rates make it difficult for them to withstand periods of heavy fishing pressure.

Vulnerability Due to Behavior

Commercial fishermen have little difficulty locating and capturing paddlefish because of their schooling tendencies, predictable spawning runs, and distribution during winter months.

Paddlefish do not form schools in which movements of individuals are coordinated. However, although direct evidence is lacking, there are indications that paddlefish do form loose schools throughout the year for most of their lives. Rosen[429] located paddlefish concentrations in slackwater areas downstream from sandbars and other current obstructions in an unchannelized section of the Missouri River, South Dakota, downstream from Gavins Point Dam. In a channelized section farther downstream, paddlefish were found in oxbow lakes and in reduced current areas behind wing dams and in inlets. In these instances, paddlefish concentrations appeared related to selection of quiet water habitat or areas of reduced water velocity.

However, in lakes and reservoirs, paddlefish distribution and behavior indicates a greater tendency toward schooling. Hevel and Alexander[246] used a midwater trawl in conjunction with a chart-recording echolocator to find and sample mixed-age (primarily young-of-year and yearling) concentrations of paddlefish in Old Hickory Reservoir, Tennessee. The largest group they found did not move more than 1 mi. from February through April. Wagner[536] thought paddlefish moved throughout Lake Pepin, Wisconsin, in large schools, because large numbers were caught irregularly in a seine 1,800 ft. long. Irregular harvest of large numbers of paddlefish was also observed in the snagline and gill net fishery on Tennessee River impoundments (Bryan[76]). Whether paddlefish concentrate in response to environmental stimuli or because of innate behavior, it is clear that large numbers may be located and caught at one time.

During winter, paddlefish form schools and consistently select deep water areas. Rosen[429] reported increased density in deeper channel areas of the Missouri River during winter. Hevel and Alexander[246] collected paddlefish from the old river channel during winter in Old Hickory Reservoir. Most paddlefish harvested by commercial fishermen in the Tennessee and Cumberland rivers during winter are captured in river channel areas. Although paddlefish are not harvested as easily or in as large numbers during winter as spring, their selection of deep channel areas in winter makes it possible for commercial fishermen to locate concentrations.

Paddlefish move to spawning areas in large streams during periods of high flow as water temperatures rise above 50 F (Purkett[407]; Pasch[393]; Russell[446]). Adult paddlefish are most readily caught just before and during these spawning runs, because they are more concentrated and easier to locate than during the rest of the year. During this time, floating gill nets fished near the surface at night are particularly effective.

Paddlefish are very responsive to river flows during spawning runs (Southall[477]; Russell[446]). On unregulated rivers paddlefish move upstream as long as flows remain high. If flows diminish, the fish move back downstream to deep water. In systems where upstream movement is blocked by dams, paddlefish concentrate below the dams during periods of significant water release. When water release is curtailed, they move downstream a few miles, usually to deep pools.

Commercial fishermen often take advantage of this behavior. On unregulated rivers, they follow paddlefish concentrations upstream or downstream, depending on flows. Downstream from hydroelectric dams, flows decrease during non-generation periods to fishable levels, making large numbers of fish susceptible to netting.

Paddlefish are harvested primarily with large-mesh gill nets; 6 to 7-in. bar-mesh nets appear to be most effective and are typically used by commercial fishermen (Alexander and Peterson[9]). Snag lines are often used when netting is illegal. Snagging by rod and reel is effective immediately downstream of dams during water release periods that preclude the use of nets and snaglines.

Eggs reach their prime in late winter and early spring, before becoming fully ripe (Jarvis 1943), and fishing is most intense during these seasons.

Vulnerability Due to Recruitment and Growth Rates

Paddlefish do not reach sexual maturity until they are 7-12 years old, and may live up to 30 years (Russell[446]). This slow maturation means that paddlefish populations have lower turnover rates than most freshwater fishes. Fish species with low natural turnover rates will not sustain harvest levels as high as those with high turnover rates (Hackney and Minns 1974). Thus, paddlefish populations are particularly vulnerable to depletion during periods of high market demand.

No exploitation rates are available for paddlefish populations known to have been depleted by harvest. However, the exploitation rates for some fished populations have been estimated and their effects monitored. Combs[112] reported that annual exploitation rates up to 19% did not have detrimental impacts on sport and mixed sport/commercial paddlefish fisheries. Purkett[409] attributed declining catch rates in the Osage River, Missouri, paddlefish sport fishery to annual exploitation rates of 7-16% during 1958-1961. Continued monitoring of the Osage River fishery revealed that an average annual exploitation rate of 11% did not affect population structure or detrimentally impact the fishery (Combs[112]).

Gengerke[195] noted a decrease in mean size and age of paddlefish harvested from Pool 13 of the Mississippi River during 1975-1978. Combined annual sport and commercial exploitation rates ranged from 11-22% during this study, but were not considered excessive.

What constitutes excessive exploitation depends on each fishery manager's perspective and goals. Generally, a large increase in harvest will result in a decrease in average size and numbers caught. Changes in size or age structure and abundance are acceptable as long as fisheries do not collapse. However, paddlefish populations exploited for roe have often collapsed; that is, they became uneconomical to harvest because of low catch rates.

Case Histories

The following examples demonstrate what commonly occurs when paddlefish are specifically sought or are an important component in a commercial fishery. In all of these examples, unfortunately, only harvest data were gathered, and we do not know associated mortality and exploitation rates.

As we noted previously, during the first two decades of this century several biologists expressed concern about rapid depletion of paddlefish populations. Stockard[489], Alexander[10], and Coker[104] reported that lakes in Mississippi, Louisiana, and Wisconsin were depopulated in a few years by intensive seine fisheries for paddlefish. However, fishing effort and harvest in individual bodies of water during periods of depletion were not reported. Only regional or national harvest statistics are available.

In 1940, the Tennessee Valley Authority (TVA) began collecting comprehensive commercial catch data for Wheeler and Wilson reservoirs in Alabama, and Pickwick Reservoir in Alabama, Mississippi and Tennessee (Bryan and Tarzwell[78]). At that time, nets were allowed only in the Mississippi and Tennessee waters of Pickwick Reservoir; set lines were fished in Alabama waters.

Before 1941 paddlefish harvest was insignificant, less than 2% of total harvest by weight in all three reservoirs. In 1941, a snag-line fishery for paddlefish developed in Wilson Reservoir. Unfortunately, wartime gasoline shortages curtailed the TVA commercial harvest survey between March 1941 and November 1942, and systematically collected data are unavailable for this period. However, Bryan[76] estimated 714,000 lb. live weight of paddlefish were taken from Wilson and Wheeler reservoirs between August 1941 and July 1942.

The commercial harvest survey was begun again in November 1942; however, only the combined catch from all reservoirs surveyed was reported. From Nov. 1, 1942 to July 1, 1943, 503,200 lb. of paddlefish were taken from Wilson, Wheeler, and Pickwick reservoirs. Beginning July 1, 1943, Guntersville Reservoir, Alabama, the next impoundment on the Tennessee River upstream from Wheeler Reservoir, was added to the commercial harvest survey. Total paddlefish harvest estimates for these four Tennessee River impoundments from 1942 through 1954 are in Table 1.

Although paddlefish landings prior to 1943 were incompletely reported, the existing data clearly demonstrate that paddlefish catch declined rapidly from 1941 to 1944. Very abundant stocks composed of many age groups were cropped quickly with snag lines. In 1946, the Alabama legislature legalized use of nets (a more efficient method than snag lines) in TVA impoundments, and harvest increased slightly for a few years. TVA biologists noted indications of excessive exploitation, but stated: "Since these fish are now in

Table 1. *Paddlefish harvest from Guntersville, Wheeler, Wilson, and Pickwick reservoirs, Alabama, from August 1941 to July 1, 1954.*

Fiscal year (beginning July 1)	Pounds of paddlefish landed (live weight)
1942	714,000[1]
1943	503,200[2]
1944	393,000
1945	117,735
1946	114,665
1947	151,555
1948	138,675
1949	165,100
1950	143,475
1951	126,880
1952	104,850
1953	138,600
1954	118,500

[1]Catch estimated for the period August 1941-July 1942 for Wilson and Wheeler only.

[2]Catch estimated for the period November 1, 1942-July 1, 1943 for Wilson, Wheeler and Pickwick only.

high demand, but have little sale value in normal times, the overexploitation is considered desirable" (Tenneseee Valley Authority[499]). Although harvest during these years may have been excessive, an equilibrium yield much lower than in 1941 or 1942 was evidently established, and the harvest remained fairly stable into the mid-1950s.

Another example of rapid depletion of a virgin stock occurred when Norris Reservoir, a deep, 32,200 a. impoundment of the Clinch River in northeast Tennessee, was opened to contracted gill and trammel netting from November through February 1958-1961 (Carroll[97]). Paddlefish catch rate and numbers harvested declined dramatically over the three fishing seasons; 781 fish (44,658 lb.) were caught the first season, 233 (12,532 lb.) the second, and 63 (2,357 lb.) the third. Average weight of paddlefish declined by about 20 lb. over the three seasons. Catches of paddlefish and flathead catfish (*Pylodictis olivaris*), the two most valuable species, were so low the third season that the contractor requested and was granted permission to terminate fishing Dec. 15, 1960. Norris Reservoir has remained closed to commercial fishing since then.

The great reduction in catch over 3 years suggests the adult paddlefish stock in Norris Reservoir was decimated. However, Alexander and Peterson[9] estimated Norris Reservoir contained 8,772 paddlefish longer than 43.8 in. TL (average weight 36 lb.) in spring 1980. These results indicate a harvestable population and suggest that the paddlefish stock recovered over a 20-year period.

Rapid decline in commercial harvest has also been observed in Barkley Reservoir and Lake Cumberland on the Cumberland River in Kentucky. In Barkley Reservoir, 7,015 paddlefish (average weight 19.5 lb.) were taken with gill and trammel nets from November 1979 through March 1980 (Bowers[58]). Harvest declined to 3,685 fish (average weight 12.8 lb.) during the 1980-1981 season. On Lake Cumberland, 9 months of commercial netting beginning in 1968 were sufficient to reduce the paddlefish catch rate to an uneconomical level (Harned[239]). Carlson and Bonislawsky[95], however, reported harvest of 414,500 lb. of paddlefish from Lake Cumberland during 18 months of commercial fishing in 1977-1979, which demonstrated that some recovery had occurred.

Old Hickory Reservoir — an impoundment of the Cumberland River near Nashville, Tennessee — was fished heavily for paddlefish from spring 1979 to fall 1981, and many fish were harvested. (Fishing effort and harvest were not estimated, but were based on our observations.) Catch rate evidently fell to an unsatisfactory level in fall 1981, because fishing pressure decreased considerably at that time. Average live weight of paddlefish also decreased over this period, dropping from 21 lb. in 1977-1979 (unpublished data, 1983, Ronald Pasch, Biologist, Tenneseee Valley Authority, Knoxville, Tennessee) to 11.6 lb. in 1981 (unpublished data, 1983, Douglas Pelren, Fish Data Collection Biologist, Tennessee Wildlife Resources Agency, Nashville, Tennessee).

Alexander and McDonough[8] found that this reduction in adult abundance in Old Hickory Reservoir may have affected reproductive success. They had previously found a strong correlation between young-of-year (YOY) paddlefish abundance, measured by YOY impingement on water intake screens at TVA's Gallatin Steam-Electric Plant, and certain environmental variables (Cumberland River discharge and temperature) during the spawning season. During 1981 and 1982, numbers of YOY impinged were lower than expected for spawning conditions during those years. They attributed this to a reduced number of spawners.

Discussion

Commercial exploitation has been a major factor affecting paddlefish populations in several southern reservoirs. Case histories provide substantial evidence that commercial harvest can rapidly deplete adult paddlefish stocks.

For many years, the fluctuating demand for roe and flesh resulted in paddlefish fisheries going through "boom and bust" cycles. In these cycles, high demand resulted in overexploitation, followed by a rapid population decline. This decline led to decreased fishing pressure and interest in the resource. Stocks were left to recover — a process that sometimes took decades.

High roe prices often caused such overexploitation. However, some paddlefish stocks were reduced to levels producing unsatisfactory catches when roe prices were below $1/lb., as was the case with the Norris Reservoir and Lake Cumberland stocks.

A recent, dramatic increase in roe prices promises to initiate another of these cycles. With roe prices currently over $20/lb. (and many females worth over $200), commercial fishermen can afford to continue fishing depleted populations that formerly would have been abandoned as unprofitable. Improvements in equipment and technology (echolocators, floating gill nets) have increased harvest efficiency. If roe prices stay high, paddlefish fisheries in some areas may be seriously damaged.

Although we know of no case where a paddlefish population has been extirpated solely by over exploitation, we believe it unwise to reduce adult abundance to levels that will not maintain a fishery. When this occurs, population recovery may take many years, the length of time depending on fishing pressure, spawning conditions and the level to which the population was reduced.

Cushing (1968) and Larkin (1978) report that recruitment is often unrelated to adult abundance when stock levels are high enough to support fisheries. However, evidence suggests that fishing pressure can reduce adult paddlefish abundance to levels that do affect recruitment. Young-of-year production in Old Hickory Reservoir decreased following heavy exploitation in 1979, although environmental conditions were favorable for spawning (Alexander and McDonough[8]).

Even when mature fish are abundant, paddlefish reproductive success can be highly variable, and dependent on river discharge and temperature during the spawning season (Alexander and McDonough[8]). When a population is depleted, adverse environmental conditions can increase both the time required for recovery, and the probability that the remaining stock will die without reproducing successfully. Likelihood of such reproductive failure increases when river systems become compartmentalized by dams that modify flows and temperatures in the remaining spawning habitat (Unkenholz[528]).

Recovery of a depleted paddlefish population may also be affected by reduced average age of spawners. Cushing (1968) and Larkin (1978) suggest that eggs from younger spawners may be less viable than those from older and larger fish.

Recommendations

We believe commercial harvest of

paddlefish should be more stringently regulated than at present. For over 80 years, biologists have observed and sometimes documented decimation of paddlefish populations under pressure from market forces. Today the increased price of roe may encourage a level of commercial fishing which will permanently damage or destroy some fisheries. In addition, paddlefish populations must now contend with the effects of man-made disturbances such as dams, pollution and channelization. In the face of these threats, paddlefish populations need to be carefully regulated.

Biologists commonly think in terms of maximum sustained yield when considering regulation of a commercial fishery. This concept implies an annual surplus production that can be harvested at no expense to the population. In the case of paddlefish, survival of year-classes to recruitment remains unknown; therefore, we are not yet able to predict annual surplus production. It is probably low, because recruitment rates are low and because one or two year-classes may support a fishery for several years (Russell[446]).

Nevertheless, paddlefish stocks can be properly managed for sustained commercial harvest even if important aspects of paddlefish biology are poorly understood. The first step in developing such a management plan is to determine the appropriate exploitation rate for the population. To accomplish this, a population size estimate should be made and at least 2 years of controlled harvest completed. Changes in numbers, age structure, or size structure after the second fishing season may indicate whether the rate of exploitation is excessive. If this approach is impractical, the commercial catch could be monitored for indications of excessive fishing mortality such as decreasing average weight and catch per unit effort. Although exploitation rates have not been determined for many paddlefish populations, there is evidence that some stocks can maintain their structural integrity at exploitation rates of 15-20%.

Other than controlling harvest, management options include regulating waterflows below hydroelectric dams to enhance paddlefish reproduction where habitat is suitable, and artificial progagation and culture of paddlefish for stocking. Such options are expensive, but may be considered if paddlefish stocks decline to critical levels. Whatever the management techniques employed, commercial paddlefish fisheries will remain economic assets only if they are properly managed.

References

CUSHING, D. H. 1968. Fisheries biology; A study in population dynamics. University of Wisconsin Press, Madison, Wisconsin, USA. 200 pages.

HACKNEY, P. A., AND C. K. MINNS. 1974. A computer model of biomass dynamics and food competition with implications for its use in fishery management. Transactions of the American Fisheries Society 103(2):215-225.

JARVIS, N. D. 1943. Principles and methods in the canning of fishery products. United States Department of the Interior, Fish and Wildlife Service Research Report Number 7, Washington, D.C., USA. 336 pages.

LARKIN, P. A. 1978. Fisheries management — an essay for ecologists. Annual Review of Ecological Systematics 9:57-63.

Effects of Dams and Other Habitat Alterations on Paddlefish Sport Fisheries

Dennis G. Unkenholz
South Dakota Department of Game, Fish and Parks
Pierre, South Dakota 57501

Abstract

Paddlefish (*Polyodon spathula*) populations have changed throughout their range since construction of dams. In a survey of 22 states within the paddlefish's range, nine states reported a reduction in paddlefish numbers since dam construction, while four states reported an increase.

The survey results, along with a literature review, revealed the effects of dams upon paddlefish habitat and populations. Dams affect paddlefish populations in a river system by altering the natural water flow and quality, by restricting migration, and by inundating spawning areas. In some cases fisheries may be temporarily enhanced by dams when fish congregate in tailwaters during spawning runs.

The effects of dams on paddlefish populations are dramatically illustrated by the case histories of Lake Francis Case, South Dakota, and Lake of the Ozarks, Missouri. After Lake Francis Case was built, paddlefish fisheries rapidly declined in a stretch of impounded river lacking natural recruitment. In Lake of the Ozarks, the fishery was initially enhanced but later threatened by construction of a reservoir upstream.

Other habitat alterations affecting paddlefish include pollution, dredging and channelization, although the survey indicated that little is known about their influences on paddlefish populations.

Historically, paddlefish (*Polyodon spathula*) moved freely up and down the large rivers in the Mississippi River drainage in response to natural events. In recent years, however, these fish have been confined to discrete sections of river by construction of large dams, which provide hydropower, flood control and navigational benefits. These dams subject the paddlefish to regulated flows, abnormal temperatures and declining water quality.

Although paddlefish were first harvested commercially in the late 1800s, sport fishing did not develop until dam construction began in the early 1900s. The dams concentrated paddlefish in tailwater areas and made blind snagging possible. Ironically, the same structures that led to development of a sport fishery also inundated traditional spawning grounds, changed the water flow and degraded the water quality necessary for natural reproduction of these fish.

In addition to dam construction, habitat changes such as pollution, dredging and siltation have adversely affected paddlefish in recent years.

In this paper, I have gathered and analyzed current information on the effects of dams and other habitat alterations on paddlefish sport fisheries. The information was obtained from published papers, through personal communication with fisheries biologists, and from a questionnaire sent to 22 biologists or agencies in February

1983. The questionnaire is described by Gengerke[196]. Responses are summarized where appropriate. I also present two case histories — Lake Francis Case, South Dakota, and Lake of the Ozarks, Missouri — illustrating the effects of dams on paddlefish sport fisheries.

Effects of Dams on Paddlefish Sport Fisheries

Paddlefish populations have changed throughout most of their range since the construction of dams. In several instances, populations have increased or decreased in part of a state, but have remained stable in another portion. [The status of various paddlefish populations is summarized in Gengerke, Table 4, Page 27.]

Although it is unlikely that these population changes are due entirely to dams, such structures clearly have significant impacts on paddlefish fisheries. Dams may alter a population of paddlefish in a river system by changing the natural water flow and quality, by restricting the fish's natural migration, or by inundating spawning areas. These effects are documented and discussed below.

Changes in Water Flow and Quality

Seasonal success of paddlefish reproduction and sport harvest is affected by both water flow and quality (Bonislawsky[55]; Van Eeckhout[531]; Berg[42]).

Specific flows have been reported as critical for spawning and/or harvest. Gengerke[195] stated that flows of 24,300 cubic feet per second (cfs) or greater were necessary to stimulate spawning activities in the Mississippi River, Iowa. Russell[450] found that a river stage of 10 ft. or a flow of 12,000 cfs was required for spawning runs in the Osage River, Missouri. They also concluded that sport harvest in the Osage River increased during rises in river stage.

Flow velocities of 10,000-20,000 cfs were necessary to stimulate a run of paddlefish from the Grand Lake of the Cherokees up the Neosho River in Kansas (Bonislawsky[55]). Paddlefish sport harvest has been directly proportional to flows in North Dakota (Van Eeckhout[531]), Montana (Berg[42]), and Oklahoma and Kansas (Bonislawsky[55]).

Water quality has also been identified as an important parameter affecting paddlefish reproduction and movement. In a study of the limnology of the Osage River and its tributaries in Missouri, Russell[450] found that the South Grand and the Marais des Cygnes rivers are limnologically similar to the Osage River. Paddlefish reproduction has been documented in both the South Grand and Osage rivers, and paddlefish movement has been documented in the Marais des Cygnes River. Three other tributaries of the Osage River — the Pomme de Terre, the Sac, and the Marmaton rivers — were not limnologically similar to the Osage River. Little or no paddlefish use has been documented in these tributaries (Russell[450]).

Information collected from the Yellowstone River during a spring fishery suggests paddlefish making a run out of Lake Sakakawea prefer the Yellowstone River over the Missouri River. During 1973 the Yellowstone River had a higher discharge, a higher suspended load, and warmer water than the Missouri River, suggesting that those factors may affect selection of spawning sites (Rehwinkel[419]).

In short, research has shown that paddlefish reproduction is influenced by certain flows, temperatures, turbidities, and possibly other water quality parameters. This suggests that dam operations that control river flow may have negative effects on paddlefish populations.

Flows in most major rivers within the range of paddlefish distribution are now regulated by dams for navigation, flood control, power production or irrigation. These regulated flows, in many cases, do not resemble natural flows in temperature, velocity, timing or duration. Annual regulated releases from Gavins Point Dam, South Dakota, are an example. Figure 1 compares the present seasonal redistribution of flows to the unregulated flows that would occur if the Missouri River dams were not operational. The high unregulated flows of March-April represent runoff from the plains, and the peak that occurs in July represents snowmelt from the mountains. These flow fluctuations are reduced by the dams on the system. The peaks are leveled off, and the traditional low flows of August, September and October are augmented.

In addition to occurring at inappropriate times, unnatural regulated flows have decreased turbidity and altered water temperature/flow regimes. In total these changes have had negative effects on paddlefish reproduction.

Sport harvest may also be affected when flow is modified by power generation. Weekly peaking cycles at Big Bend Dam, South Dakota, produced reduced flows, usually during the weekend when power demands were low. These periods of low flow corresponded with periods of low harvest (Fig. 2). This indicated that paddlefish moved downstream in response to reduced flows and required a day or two to move back to the tailwater area once flows resumed (Friberg[187]).

Restriction of Movement

Paddlefish are capable of moving long distances either upstream or downstream in response to spawning instincts, decreasing water flows or random chance.

Extensive upstream spawning movements have been documented at Fort Peck Reservoir, Montana (Berg [42]);

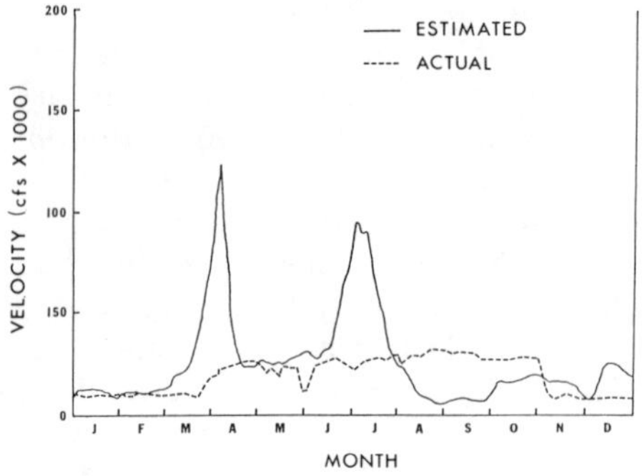

Figure 1. *Actual (regulated) and estimated (without dam) flows in the Missouri River below Gavins Point Dam during 1959.*

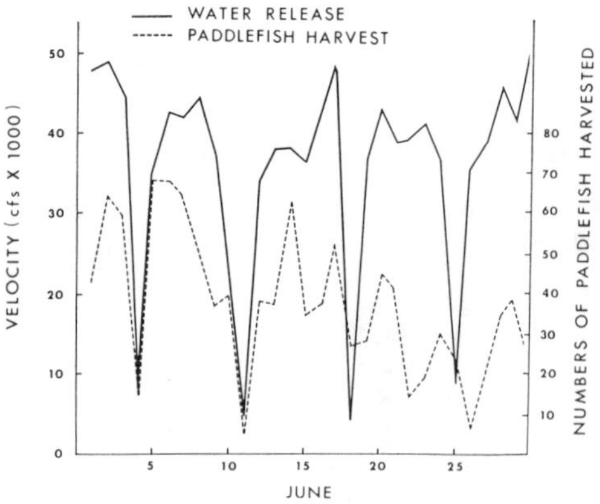

Figure 2. *Relationship between average daily water release into Big Bend Dam tailwater and estimated harvest of paddlefish, June 1972.*

Lake Sakakawea, North Dakota (Rehwinkel[419]; Van Eeckhout[531]); Grand Lake of the Cherokees, Oklahoma (Bonislawsky[55]); and Lake of the Ozarks, Missouri (Russell[450]).

Movements of paddlefish downstream from Gavins Point Dam, South Dakota, also have been documented. This dam, the lowermost on the Missouri River, blocks upstream movement of fish that are free to move downstream into the Mississippi River. Unkenholz[526] reported that approximately 12% of all tag returns from paddlefish tagged in the Gavins Point Dam tailwaters came from downstream states. Figure 3 indicates where these fish were caught. The magnitude of movement is best illustrated by a tag return just upstream from Memphis, Tennessee. That fish traveled 800 mi. down the Missouri River and 440 mi. down the Mississippi River, a total of 1,240 mi.

Figure 3. *Recapture sites for paddlefish tagged in the Gavins Point Dam tailwater in South Dakota, but recaptured in other states (each dot represents a recaptured paddlefish).*

Long-range movements have also been demonstrated in Montana, where paddlefish tagged at Intake on the Yellowstone River have been caught in Lake Sakakawea (Van Eeckhout[531]) or in the Fort Peck Dam tailwaters (Fig. 4) (Rehwinkel[419]). Paddlefish regularly interchange between the Yellowstone and Missouri rivers. These fish apparently reside in Lake Sakakawea and move upriver to spawn. They seem to prefer the Yellowstone River over the Missouri River as a spawning site (Rehwinkel[419]).

Paddlefish movement has been documented from Lake of the Ozarks, Missouri, up the Osage River into Kansas, a trip of approximately 200 mi. (personal communication, 1983, L. Kim Graham, Fisheries Research Biologist, Missouri Department of Conservation, Columbia, Missouri).

It is clear from these studies that long-range upstream and downstream movements may be important for long-term survival of the paddlefish. However, dam construction has blocked these natural movements throughout its range. What once was a population spread over the Mississippi River drainage is now a collection of isolated segments of that population locked between dams. For instance, paddlefish in South Dakota are now confined to sections of the Missouri River by the main-stem dams (Fig. 5).

Destruction of Spawning Areas

Virtually no information existed on paddlefish spawning sites until long after dam construction had begun, except in the case of the Osage River, Missouri, where paddlefish spawning habitat was well defined. It is therefore impossible to say how many spawning areas have been destroyed by dams. However, there is no doubt that destruction did occur, since the six main-stem Missouri River dams flooded hundreds of miles of Missouri River.

Destruction of spawning areas has resulted from river bed degradation below the dams, erosion of river banks, siltation, and changes in water quality resulting from operation of the mainstem dams.

Case Histories

To further illustrate the effects of dam construction on paddlefish sport fisheries, two case histories are presented. The first describes the rapid decline of the sport fishery in a stretch of impounded river lacking natural recruitment. The second describes a fishery initially enhanced by the construction of a reservoir, and later threatened by construction of a second reservoir.

Lake Francis Case in South Dakota was formed by the closure of Fort Randall Dam in 1952. Paddlefish appeared in the tailwaters immediately and the first reported harvest was in 1954 (Shields[462]). In 1956 Oahe Dam in South Dakota was completed, 192 mi. upstream (Fig. 5). A significant fishery developed in the Oahe tailwaters in 1960-61, and approximately 3,200 fish were harvested during that period and 4,400 the following year. By 1963, however, harvest had dropped to 700 fish (Fogle[175]). During the fall of 1963, Big Bend Dam was closed 85 mi. downstream from Oahe Dam (Fig. 5). The closure of Big Bend Dam corresponded with the decrease in sport harvest at Oahe Dam tailwaters.

Shortly after closure of Big Bend Dam an active fishery developed in the tailwaters. In a few years, however, it became evident that there was a lack of young individuals in the sport harvest in Lake Francis Case, and a creel survey and tag-and-release study

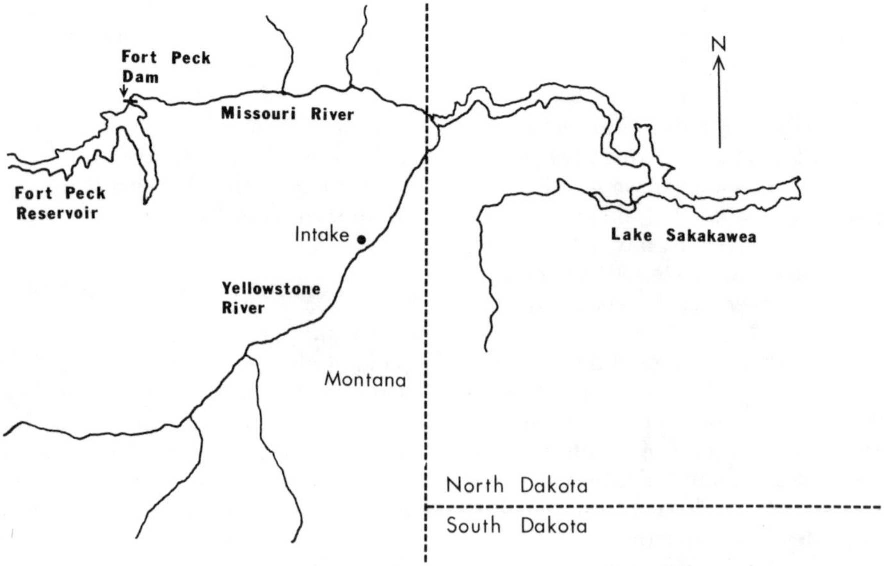

Figure 4. *Fort Peck Reservoir and Lake Sakakawea, upper Missouri River and lower Yellowstone River.*

were initiated in 1970. Sport harvest at Big Bend Dam peaked in 1971 at 2,831 fish, and declined through 1979 (Fig. 6); few fish have been caught since 1979.

The Lake Francis Case case study shows clearly what happened to a paddlefish population when migration routes were blocked and spawning areas inundated by dams. Sport harvest may increase temporarily as migrating adults concentrate in the dam tailwaters, but decline is inevitable since no new individuals are recruited to the population.

A similar situation may be developing in Missouri, where Lake of the Ozarks was formed by the construction of Bagnell Dam in 1931. This dam prevented fish in the upper Osage River from moving downstream into the Missouri River. These fish apparently prospered and increased in numbers because of the productive reservoir environment. Spawning was successful, since fish were free to move upstream and use the gravel bars of the Osage River (Purkett[407, 409]). Concentrations of paddlefish were observed by anglers and a snag fishery developed in the Osage River in the early 1950s. Sport harvest occurred upstream as far as the Osceola Dam (Fig. 7).

Paddlefish sport harvest from the Osage River has varied, however. From 1973 to 1979, harvest ranged from 1,910 to 5,430 with a mean annual catch of 2,720 fish (Russell[450]). Sport harvest occurred mainly during spawning runs from Lake of the Ozarks upstream into the Osage River (Purkett[409]; Russell[450]).

In 1977, the construction of Harry S. Truman Dam, 93 mi. above Bagnell Dam on the Osage River, trapped large numbers of paddlefish in Lake of the

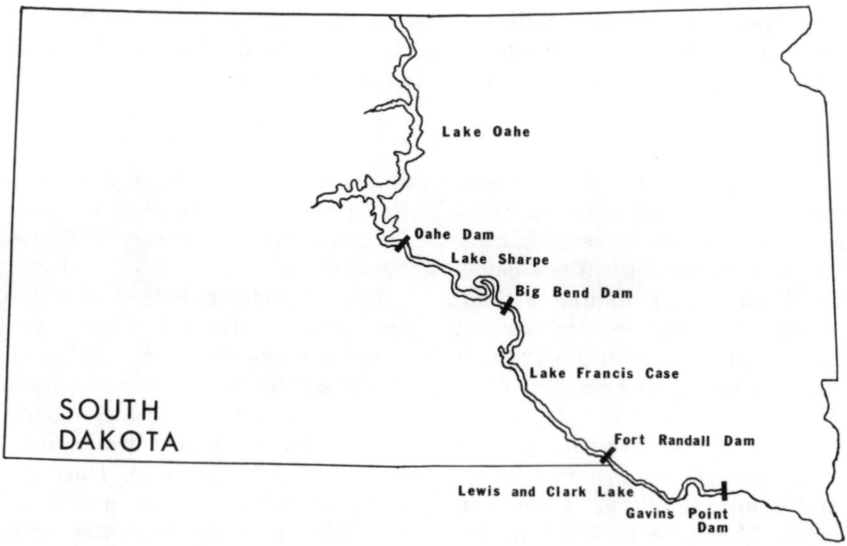

Figure 5. *Missouri River reservoirs in South Dakota.*

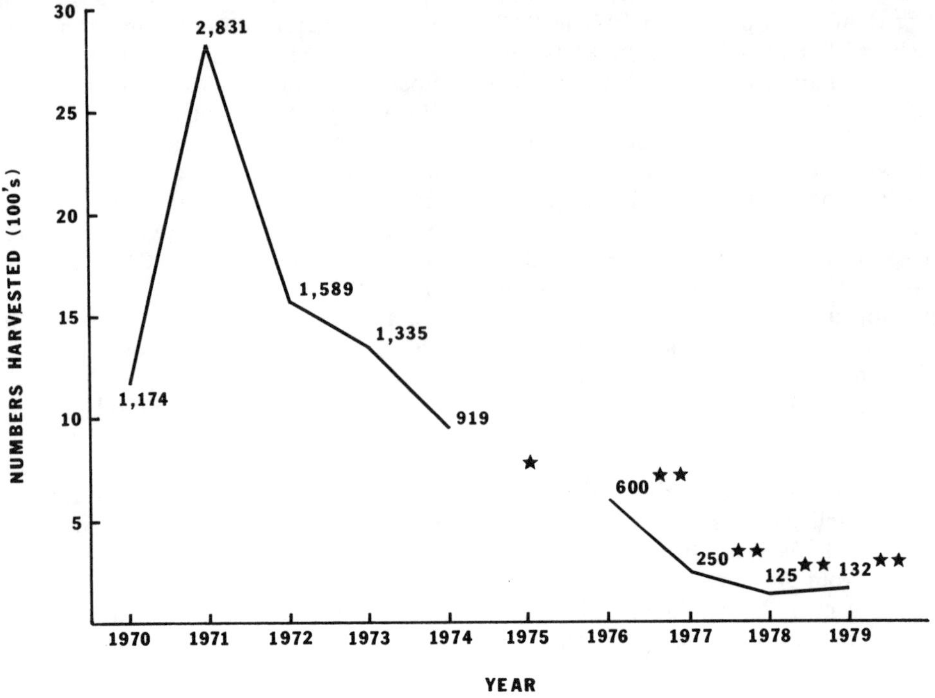

Figure 6. *Sport harvest of paddlefish below Big Bend Dam, 1970-1979. *Data unavailable. **Number harvested based on Corps of Engineer's counts of paddlefish heads deposited in garbage cans near the boat ramp.*

Ozarks and eliminated their upstream spawning movement (Russell[450]). Sport harvest of these fish increased, however, due to their concentrations below Truman Dam. Sport fishing regulations designed to prevent overharvest have been implemented in the Truman Dam tailwaters (Russell[450]; Graham[217]).

Adult paddlefish are present in unknown numbers in Truman Reservoir, and young paddlefish have been introduced. Biologists in Missouri are watching for a run of paddlefish in the Marais des Cygnes River upstream into Kansas. Paddlefish numbers in eastern Kansas could be increased if a spawning run developed from Truman Reservoir up the Osage and Marais des Cygnes rivers (Russell[450]; Bonislawsky[55]).

Loss of paddlefish due directly to the operation of Truman Dam has also been documented. The facility was designed to function as a pumped storage system. During the pumping mode, paddlefish are drawn into the turbines and mutilated. During high spillway releases, fish are damaged and killed by turbulent water (personal communication, 1983, Tom Russell, Fisheries Research Supervisor, Missouri Department of Conservation, Columbia, Missouri).

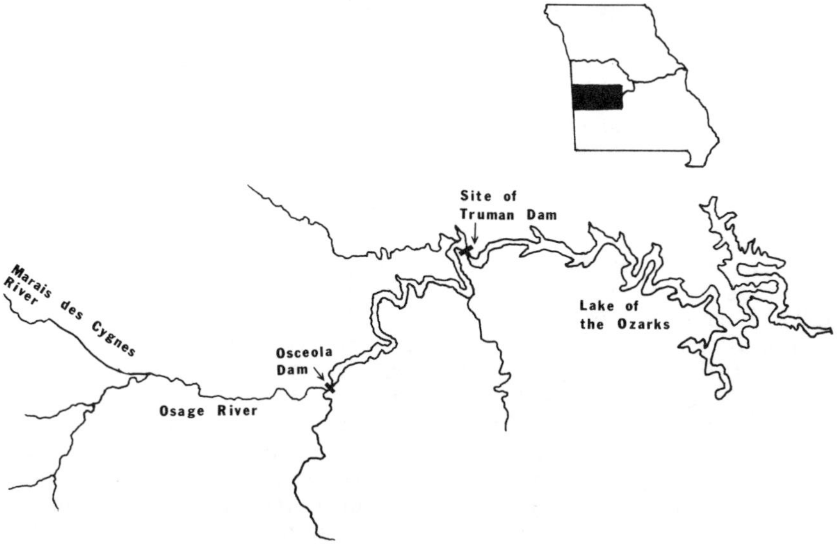

Figure 7. *Lake of the Ozarks — Osage River complex.*

Effects of Other Habitat Alterations on Paddlefish

Little is known about the effects of pollution, dredging, channelization and other environmental changes on paddlefish populations. Most survey responses were based on casual observations, limited netting, or harvest data. Gross pollution in the upper Mississippi River was suggested as a possible cause of decline in paddlefish populations there (Eddy and Underhill[144]). Pesticides and decreased water quality due to siltation may have negatively affected paddlefish populations in Mississippi (personal communication, 1983, C.A. Schultz, Planning and Programming Supervisor, Mississippi Game and Fish Commission, Aberdeen, Mississippi). Loss of spawning areas due to dam construction, dredging, and channelization may have contributed to the decline of paddlefish in Texas (personal communication, 1983, Alan G. Wegner, Regional Director, Inland Fisheries, Texas Parks and Wildlife Department, Austin, Texas).

No state indicated sport fishing had negatively affected paddlefish populations.

Conclusions

The net effect of dams on paddlefish sport fisheries seems to be negative. The bittersweet reality is that, while the construction of dams allowed the formation of sport fisheries, in many instances a decline in paddlefish reproduction has been a result of those same dams. Other habitat changes, including pollution, siltation and dredging, have also been damaging. Future demands on large river resources in the native paddlefish range can mean only a greater negative effect on paddlefish sport fisheries.

An Overview of Current Management Practices for Paddlefish Fisheries

Allen A. Elser[1]
Montana Department of Fish, Wildlife and Parks
Helena, Montana 59620

Abstract

Current paddlefish (*Polyodon spathula*) management practices include those which regulate fishing, protect paddlefish habitat, and encourage public support for paddlefish conservation.

To determine current management philosophies, fishing regulations from 22 states within the present range of the paddlefish were reviewed. Of these, 11 states consider paddlefish as non-game or non-sport fish; 7 states identify the fish as game or sport fish; and 4 states totally protect paddlefish. Of 18 states that allow paddlefish harvest, 11 permit commercial fishing. Regulations (sport or commercial) are generally designed to maximize long-term harvest by restricting efficiency of fishing gear, limiting total catch, or protecting fish in locations where they are particularly vulnerable. The cornerstone of paddlefish protection, however, is habitat preservation. Paddlefish have evolved as part of an aquatic system; therefore, it is this system that we must protect.

While key habitat has been lost in some areas (e.g. the Osage River), other aquatic habitat has been saved (Upper Missouri River). In the lower Yellowstone River in Montana, instream flows needed for paddlefish migration, spawning and egg incubation were assured through legislation. Private organizations such as Defenders of Wildlife and local groups such as rod and gun clubs and chambers of commerce are becoming more aware of, and vocal about, paddlefish management. Their involvement can have either positive or negative effects on paddlefish management.

Two avenues must be explored in future paddlefish management: 1) the establishment of consistent regulations based on regional characteristics of paddlefish populations, rather than on political boundaries; and 2) an increase in cooperation among states in protecting paddlefish habitat. If paddlefish are to survive, protection of the system of which they are a part must become paramount in management decisions.

The first primitive attempt to allocate fishery resources was probably made by our early ancestors. We can assume that the strongest and most aggressive of these prehistoric beings simply took whatever fish they needed, and defended their catches and fishing territories from competitors. This initial "brute force" management, somewhat refined by civilization, has continued to influence decision-making in fisheries resource allocation.

Throughout the early years of modern fisheries management, regulations limiting fish harvest played an important management role. Increasingly restrictive regulations regarding seasons, creel limits and methods of

[1]Present Address: Montana Department of Fish, Wildlife and Parks, P.O. Box 67, Kalispell, Montana 59901

taking fish were relied upon to slow the observed, or at least perceived, decline in recreational angling success. Commercial fishing has been controlled by legislation and/or regulation in most inland waters (other than the Great Lakes and big rivers) since the middle of the 20th century.

As recreational and commercial fisheries became more and more regulated, highly charged confrontations between competing user groups became common. Partly because of this pressure, the era of habitat management began, with efforts aimed at pollution abatement, streamflow allocation and critical review of impoundments. Attempts were begun to protect habitats still considered good, to enhance marginal ones and to mitigate those lost.

Current paddlefish (*Polyodon spathula*) management employs several approaches to reach its goals. Management practices include those which regulate harvest, protect fish habitat and encourage public support for paddlefish conservation. Balancing these approaches to produce the greatest harvestable surplus for all user groups is the bottom line in fisheries management.

Current Management Tactics: Regulation of Harvest

The purpose of any fishing regulation, in a strict conservation sense, is to provide greater harvest of fish, either in volume or value. Most regulations are based on the premise that fewer fish caught now will mean more available for the future. Regulations are thus designed to: 1) allow for survival of a fairly large number of breeding adults; 2) protect smaller fish so they can reach maturity; and 3) protect fish during spawning seasons.

To determine current philosophies in paddlefish management, we reviewed fishing regulations from the 22 states that are within the fish's present range (Gengerke[195]). Eleven states consider paddlefish to be non-game or non-sport. Except for North Dakota, these states are clustered in the central and southern portions of the fish's range. Paddlefish are considered game or sport fish in seven states which are mostly in the northern area of the fish's range (Fig. 1). Four states — Minnesota, Ohio, Texas, and Wisconsin — totally protect the paddlefish.

Of the 18 states that allow paddlefish harvest, 11 permit commercial fishing. As is the case with sport/non-sport designations, the more conservative regulations (non-commercial) are in headwater or upper basin states, with the exception of North Dakota. With the exception of Missouri, states that allow commercial fishing also consider paddlefish a non-sport species.

Regulations, whether sport or commercial, are generally designed to maximize overall harvest by restricting efficiency of the individual fishing unit. If we regulate the number of hooks or the length of the snag pole the angler can use, we in theory restrict his efficiency.

Limitations placed on commercial gear probably have more impact on commercial fishermen than those placed on sport fishing tackle have on sportsmen. Commercial gear restrictions are established to protect certain segments of the paddlefish population; most often these regulations limit gill net mesh size.

Bag or creel limits establish an upper limit of individual success. Managers rely on creel limits to ensure a "fair distribution" of the harvest among participants. (Here another realm is entered, that of psychology: it always sounds better to be able to say, "Yup, limited out the other day.") Perhaps spoonbill fishing is more satisfying in states with conservative regulations such as South Dakota and Nebraska than in liberal states such as Kentucky and Arkansas.)

While daily bag limits are imposed in most states, two currently use quota systems. Montana limits an individual's annual take, while South Dakota limits the annual harvest within specific management areas.

At a time when trout fishermen are being asked to release their fish, paddlefish snaggers are told to keep theirs. Such a no-release policy helps prevent additional mortality of released fish. Currently 10 states allow anglers an opportunity to select their catch and release the undersized fish, a practice called high-grading.

A fisherman's ability to catch fish also can be limited by restricting fishing time. Limitations can be established on a daily or seasonal basis. Pressures build each year in Montana to close night fishing, since harvest appears to be more successful after dark. Supporters of this policy also point out that violations are more difficult to observe at night.

Efficiency also can be influenced by establishing sanctuaries. If we know of places where paddlefish concentrate, such as spawning grounds and migration corridors, we can prohibit sport and/or commercial harvest in these areas when necessary to protect that segment of the population.

Currently, the paddlefish is totally protected in Minnesota, Ohio, Texas, and Wisconsin, while Montana recognizes it as a species of special concern. Since most protective classifications

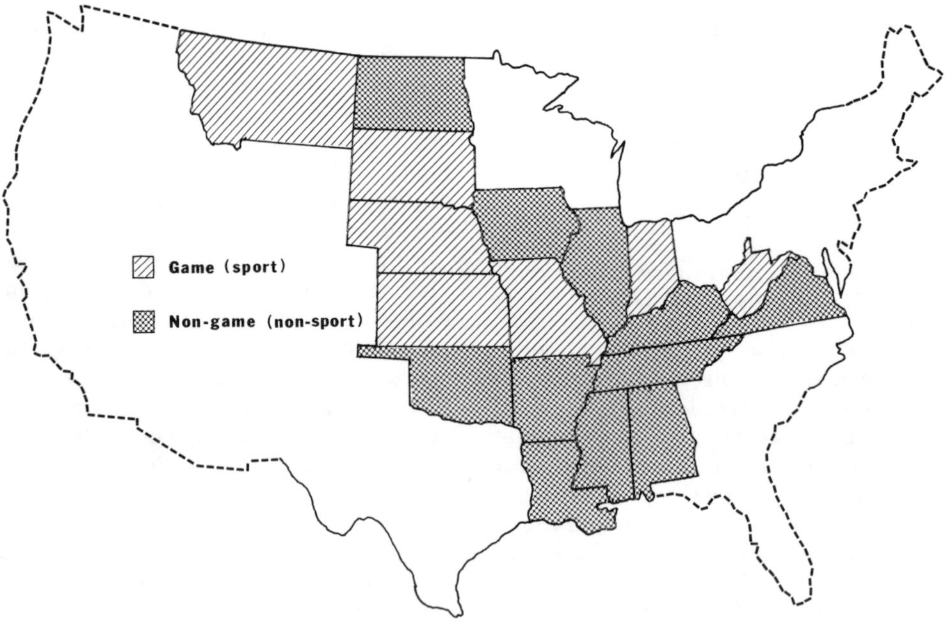

Figure 1. *Classification of paddlefish by states as either game (sport) or non-game (non-sport).*

(such as threatened or endangered) restrict harvest, and since this is neither necessary nor desirable in most areas, Carlson and Bonislawsky[95] have advised against placing paddlefish into a single protective classification over a wide area. Rather, they suggest that protective classifications be applied state-by-state or in relatively small regions. Montana has established three classes of concern for fish species. The paddlefish is considered a Class A species, meaning that there are limited numbers and/or limited habitat both in Montana and elsewhere in North America, and that elimination from Montana would be a significant loss to the gene pool of the species.

Current Management Tactics: Habitat Protection

As biologists, we know that the cornerstone of paddlefish conservation — or the conservation of any species — is habitat preservation. But we sometimes forget that, in order to protect paddlefish habitat, we must protect the entire aquatic system.

Paddlefish did not evolve in isolation; they have always shared habitat with a myriad of other species. In other words, they are part of a river system — a system we've tacked a name on, such as Missouri, Osage or Yellowstone. While we may be able to identify certain gravel bars as important to paddlefish reproduction, we must not lose sight of the importance of the rest of the system to the survival of the paddlefish and other species.

In recent years, there have been both failures and successes in protecting river systems where paddlefish are found.

When the Harry S. Truman Dam was proposed for the Osage River in Missouri, many biologists knew that the site Mr. Purkett had discovered to be an important paddlefish spawning area would be flooded. Russell[450] predicted a decline in the Osage River paddlefish population and fishery after dam closure. Biologists and concerned citizens fought a gallant battle, but the gates were closed in 1977. Now only time will tell the accuracy of that prediction.

On the other hand, not all battles have been lost. When the U.S. Army Corps of Engineers proposed High Cow Dam on the upper Missouri River in Montana in the late 1950s, biologists were uncertain that paddlefish existed in that reach of the river. In this case there was a river system to be protected, and the system itself, rather than one or two species of fish, was the key to the battle. Many user conflicts arose but finally both the river system and the paddlefish were protected. This section of the river has since become an important fishery, increasing the diversity of fishing in the Missouri River.

In the Yellowstone River Basin of Montana, we were allowed through state law to reserve a portion of the flows of that river as critical habitat for fish. Flow needs for paddlefish passage, spawning and incubation were identified in the lower Yellowstone Basin, and became an important part of a document that would ultimately protect the natural flow of the Yellowstone River. When the Montana Board of Natural Resources signed the order issuing the Yellowstone Flow Reservation on Dec. 15, 1978, they achieved a first: legal protection of water remaining in the channel for fish and wildlife use. This water right presently protects not only paddlefish, but Yellowstone cutthroat trout (*Salmo clarki*), shovelnose sturgeon (*Scaphirhynchus platorynchus*), Canada geese, whitetail deer, and the integrity of islands and riparian zones. And who knows — perhaps because the paddlefish protected the system, the system will protect other species as yet unrecognized.

Even when habitat is destroyed, the loss can sometimes be lessened through mitigation. While paddlefish losses due to Truman Dam have been high, efforts to mitigate these losses are in progress. To replace paddlefish reproduction lost in the Osage River, Missouri is negotiating a lump-sum payment to build a paddlefish hatchery, expand existing facilities or purchase young fish from a commercial dealer. The Corps of Engineers is also developing and testing a screening system to keep fish out of pumps during pump-back. If a successful system can't be developed, then pump-back will not be allowed. This protects not only paddlefish, but other important Osage River fish as well.

Below Fort Randall Dam, a Missouri River mainstem dam in South Dakota, improved flow releases resulted from cooperation between the state and the Corps of Engineers. Biologists found that peaking flows during critical spawning seasons for walleye (*Stizostedion vitreum*), sauger (*Stizostedion canadense*), paddlefish and smallmouth bass (*Micropterus dolomieui*) were destructive to the fishery. During peaking times, flows would drop to zero at night and then bounce to full capacity during the day. Mean flows looked good, but minimums were harming the fisheries. A minimum instantaneous flow of 15,000 to 20,000 cubic feet per second (cfs) was recommended and accepted by the Corps of Engineers (personal communication, 1983, Dennis G. Unkenholz, Missouri River Coordinator, South Dakota Department of Game, Fish and Parks, Pierre, South Dakota). Through better understanding of the system, the fishery was protected.

The Role of Private Organizations

The value and impact of public involvement in management efforts is increasing. The environmental movement of the 1970s produced new ecological awareness in the public; people are beginning to understand that their "neck of the woods" is a small part of the whole. Pressures generated by sportsmen's groups can bring about new appreciation of problems by government agencies. Sometimes these pressures make things happen, such as improved flow releases at Fort Randall Dam, a paddlefish hatchery on the Osage River, or instream flow protection on the Yellowstone River.

Involvement can be positive, as was the support given by Defenders of Wildlife to Montana's Department of Fish, Wildlife and Parks during the "battle of the golden eggs" in 1980. The state chapter of Defenders aligned with the Department's position prohibiting commercial exploitation of paddlefish eggs for caviar. Public involvement can also be negative, as demonstrated by a money-making fish derby at Intake, Montana, on the Yellowstone River, sponsored by a well-meaning chamber of commerce. The paddlefish season at this location approaches insanity on its best day, so imagine the hysteria resulting from hundreds of additional snaggers seeking the $500 first prize!

Tremendous success can result from public involvement, as demonstrated by Walleyes Unlimited of Montana. Less than a year old, it is the fastest growing sportsmen's group in the state, with over 3,000 members. In contrast, the International Spoonbill Society failed to gain enough support to continue.

What's in the Future?

Rational management of paddlefish as a resource should be the major goal for the future. But management is complicated by the wide geographical distribution of the paddlefish, by its differing status in each state, and by varying levels of concern over the species' welfare.

Destruction of paddlefish habitat has been widespread throughout its range. Construction of dams has impeded free movement of paddlefish within some rivers, although some populations have flourished within the restricted habitat as long as their reproductive requirements were met. Decreased freedom of movement isolates populations within states, and makes short-sighted management more likely.

Interstate movement complicates management, but offers fantastic opportunities for a regional approach. Communication between states sharing populations has been good, since limits, seasons and gear restrictions are similar. But few agreements on paddlefish habitat protection have been reached. Further cooperation is essential to evaluate the impacts of a commercial fishery on paddlefish populations, to assess the relationship of paddlefish to river systems, and to increase public awareness of the paddlefish and its problems.

The Role of Regulations in Managing Paddlefish Populations

David L. Combs[1]

Oklahoma Department of Wildlife Conservation
Porter, Oklahoma 74454

Abstract

Attempts to manage paddlefish (*Polyodon spathula*) populations through harvest regulations are relatively recent, despite a long history of sport and commercial harvests. To assess the status and role of such regulations in managing paddlefish resources, I reviewed the current regulations imposed on sport and commercial fisheries in the 22 states where paddlefish still occur.

How a regulating agency classifies a species generally determines the management philosophy. Paddlefish are classified as a non-game fish by 11 states, a game fish by seven states, and a protected species by four states. Seventeen states report sport fishing for paddlefish, and 11 of these impose regulations on the harvest. All 11 use creel limits, six use harvest seasons, six prohibit high-grading, and one state uses sanctuaries and closures. Commercial fishing is permitted by 11 states, seven of which impose regulations on the harvest. Size limits are used by three states, area and lake closures by three states, seasons by two states, and gear restrictions and harvest quotas by one state each.

The reported objectives of regulations for both sport and commercial fisheries are to regulate harvest, protect brood stocks or protect populations. However, of the 14 states regulating paddlefish harvests, only four feel that these objectives have been met. The rest concede that effects of regulations on paddlefish populations are unknown. Further investigations are needed to determine whether current regulations or some combination of new ones will best meet the management objectives set for our paddlefish fisheries.

Fisheries have been regulated on the basis of politics, social presssure, gear competition, prejudice, whim, and sometimes biology (Everhart and Youngs 1981). Despite past uses and misuses, regulations are regarded as one of three traditional techniques employed in effective fisheries management, and as the most useful technique for sustaining balance in fish populations (Anderson 1975). A regulation can be designed to manage the angler, the fish population, or both. At present, most resource agencies have relatively little information on the status of paddlefish (*Polyodon spathula*) populations in their respective states, so regulations usually focus on management of the anglers.

Regulation of paddlefish fisheries, both sport and commercial, is a relatively recent development. Paddlefish resources were overlooked and unregulated for many years, because early investigators thought the fish had little sport or food value (Hussakof[259]). Commercial harvests of paddlefish were first documented in 1894 (Smith[469]), and began to increase around the turn of the century to

[1]Present Address: U.S. Army Corps of Engineers, P.O. Box 619, Tulsa, Oklahoma 74121

augment the declining lake sturgeon (*Acipenser fulvescens*) caviar industry (Hussakof[259]). By the early 1900s, researchers recognized that commercial exploitation could deplete paddlefish stocks, and that paddlefish were fast disappearing due to lack of protection (Alexander[10]; Stockard[489]).

During this period of stock depletion, only Louisiana recognized the need for regulations to protect the fishery. In 1914, Louisiana established a 6-month closed season and a 12-in. minimum length to regulate harvest of paddlefish. Nearly 25 years elapsed before other state agencies established regulations to protect paddlefish brood stock from commercial overexploitation.

Regulations on paddlefish sport fisheries also developed gradually as concern grew over sport harvests. The first record of a paddlefish sport harvest regulation that I have found was on the Osage River, Missouri, in 1951 (Missouri Department of Conservation 1951). Only in recent years have other states with paddlefish sport fisheries restricted sport harvest to prevent overexploitation.

This paper reviews existing state regulations imposed on sport and commercial paddlefish fisheries and examines their role in managing paddlefish resources. Only those regulations specifically imposed on paddlefish populations and harvests are included. Questionnaires and telephone interviews were used to gather information from fisheries personnel in all 22 states within the current range of the paddlefish.

Classification of Paddlefish

State agencies generally base their fisheries management strategies upon how they classify any particular species. The 22 states I surveyed show regional differences in classification and management philosophies, primarily between southeastern and western states (Table 1). Paddlefish are classified as a non-game species by 11 mostly southern and eastern states, while seven states along the western range classify paddlefish as game fish. These 18 states presently allow harvests (Fig. 1), and 14 of them impose harvest regulations. Texas, Wisconsin, Minnesota and Ohio have assigned paddlefish a protected status and prohibit any harvest.

Sport Fish Regulations

Paddlefish sport fishing is permitted by 17 states (Fig. 1). All of these states have sport fishing regulations except Alabama, Arkansas, Indiana, Mississippi, Virginia and West Virginia. Five types of regulations are presently used to manage paddlefish sport fisheries. Ranked in order of use below, these are: creel limits, seasons, prohibition of high-grading, sanctuaries and closures.

Creel Limits

All 11 states regulating paddlefish sport harvests use creel limits. These vary widely, ranging from a liberal limit of 15 fish per day in Kentucky to a restrictive limit of two fish per season on the Yellowstone River in Montana. The most widely used creel limit is two fish per day, set by seven of the 11 states (Table 2).

Nebraska, South Dakota and Montana have creel limits that vary according to location of the fishery within the state. Nebraska and South Dakota have each set different limits above and below Gavins Point Dam on the Missouri River. Montana has set different limits in the Missouri River (one fish per day) and the Yellowstone River (two fish per season). Presumably, the different limits reflect the paddlefish populations available for harvest at these locations, and the fishing pressure placed upon these populations.

States that share boundary waters in the western range of the paddlefish use identical creel limits for those waters. States along the Missouri River below Gavins Point Dam (Missouri, Kansas, Nebraska, South Dakota and Iowa) have a creel limit of two fish per day. States above Gavins Point Dam (Montana, Nebraska, North Dakota, South Dakota) report a limit of one fish per day. A creel limit of two fish per day is also used by states along the upper Mississippi River (Missouri, Illinois, Iowa). Most states indicate that during recent years limits have become more restrictive as angling pressure has increased.

Possession limits range from the daily creel limit to three times the daily creel limit. Illinois and Kentucky report no possession limits.

Seasons

Six state agencies along the western range of the paddlefish use snagging seasons in management of the sport fisheries (Table 3). Seasons vary widely between states, ranging from 62 days in Kansas to 182 days in Nebraska and South Dakota. Missouri is the only state with spring and fall seasons. Their spring season is split: 47 days for Lake of the Ozarks, and 62 days for the remainder of the state. Missouri's 47-day spring season is actually the shortest season reported. Nebraska and South Dakota have identical 182-day seasons on the Missouri River above Gavins Point Dam and 152 days below the dam.

Although lengths of seasons vary, every spring season encompasses at least a portion of the spawning period. This scheduling makes paddlefish more accessible to anglers.

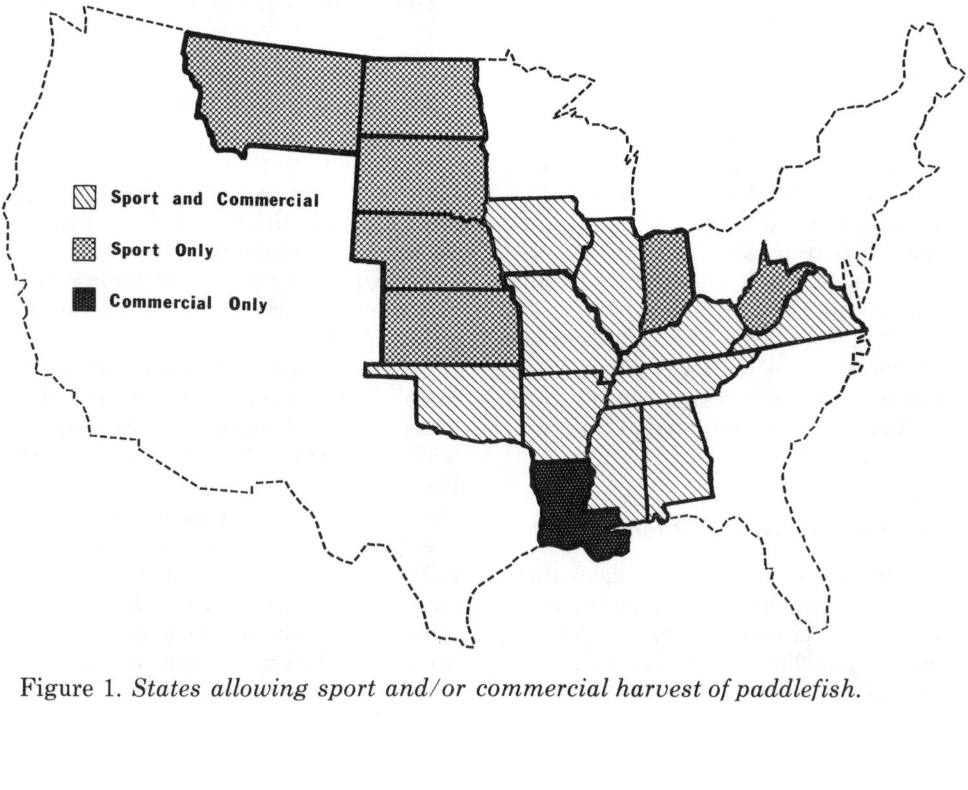

Figure 1. *States allowing sport and/or commercial harvest of paddlefish.*

Table 1. *Classification of paddlefish and presence or absence of harvest regulations.*

State	Classification: Game	Non-Game	Protected	Regulations
Alabama		X		No
Arkansas		X		Yes
Illinois		X		Yes
Indiana	X			No
Iowa		X		Yes
Kansas	X			Yes
Kentucky		X		Yes
Louisiana		X		Yes
Minnesota			X	—[1]
Mississippi		X		Yes
Missouri	X			Yes
Montana	X			Yes
Nebraska	X			Yes
North Dakota		X		Yes
Ohio			X	—[1]
Oklahoma		X		Yes
South Dakota	X			Yes
Tennessee		X		Yes
Texas			X	—[1]
Virginia		X		No
West Virginia	X			No
Wisconsin			X	—[1]
TOTALS	7	11	4	

[1]Complete protection.

Prohibition of High-Grading

Six of the 11 states that regulate harvest prohibit the practice of high-grading, which is removing and releasing smaller fish from the creel so that larger fish can be caught without exceeding the limit. Mortalities have been reported for released paddlefish with snagging wounds (Friberg[188]). Identical high-grading regulations are also used by states along the Missouri River (Montana, North and South Dakota, Nebraska).

Sanctuaries and Closures

Missouri is the only state that uses sanctuaries to protect paddlefish and prevent sport fish harvest in specific areas. Many states prohibit snagging in tailwaters, but recent regulations in Missouri represent the first use of a sanctuary specifically for paddlefish. Missouri's sanctuary is a 3.5-mi. section of the Osage River immediately below Harry S. Truman Dam, an area where paddlefish concentrate (Graham[217]).

Missouri is also the only state to use closures specifically to protect paddlefish. Truman Reservoir was closed to paddlefish sport harvest in 1977 to allow development of a paddlefish population from stocked fish that had been hatchery-reared (Graham[217]).

Commercial Fishing Regulations

Commercial fishing for paddlefish is permitted by 11 of the 22 states within the current range of the paddlefish (Fig. 1). Seven of the 11 impose harvest regulations. Five types of regulations presently being used to manage paddlefish commercial harvests are: size limits, seasons, closures, quotas and gear restrictions.

Size Limits

Minimum size limits have historically been used to protect fish until they are sexually mature, thereby increasing potential yields (Everhart and Youngs 1981). Minimum size limits are among the oldest regulations recorded for paddlefish commercial fisheries. Size limits were first used by Louisiana in 1914, and later imposed by Arkansas and Mississippi around 1940. Current regulations are somewhat comparable among these three states, with a 32-in. minimum size limit in Arkansas and Mississippi, and a 15-lb. limit in Louisiana.

Seasons

Seasons, like size limits, were among the first regulations imposed on paddlefish commercial fisheries to prevent overexploitation. Today, only Kentucky and Oklahoma use harvest seasons, of 152 and 304 days, respectively. Their seasons are closed during April and May, apparently to protect paddlefish during spawning migrations.

Closures

Kentucky, Oklahoma and Tennessee currently use commercial fishing closures to protect paddlefish populations. Kentucky has closed Cumberland Lake to all commercial netting, and Tennessee closes specific areas of Old Hickory Reservoir. Oklahoma uses both area and complete reservoir closures to prevent commercial harvests on several reservoirs.

Quotas

Harvest quotas have been used for many years in saltwater commercial fisheries but have been largely ignored

Table 2. *Creel and possession limits for paddlefish sport fisheries.*

State	Creel	Possession
Illinois	2/day	None
Iowa	2/day	4
Kansas	2/day	6
Kentucky	15/day	None
Missouri	2/day	2
Montana (Missouri River)	1/day	2
(Yellowstone River)	2/season	2
Nebraska (Above Gavins Point Dam)	1/day	1
(Below Gavins Point Dam)	2/day	2
North Dakota	1/day	2
Oklahoma	3/day	6
South Dakota (Above Gavins Point Dam)	1/day	1
(Below Gavins Point Dam)	2/day	2
Tennessee	2/day	4

by managers of inland fisheries. One exception occurs in North Dakota, where since 1974 a commercial harvest quota has permitted an annual harvest of up to 1,000 lb. of paddlefish during commercial fishing. This regulation provides for incidental harvest of paddlefish but prevents large-scale commercial exploitation. It represents the first use of a quota regulation for commercial harvest of paddlefish.

Gear Restriction

Gear restrictions are commonly placed on types of netting, numbers of nets, and mesh sizes to regulate commercial fishing. The only gear restriction placed specifically on a commercial paddlefish fishery is a maximum bar mesh size of 4.5 in. imposed by Kentucky. This regulation prevents excessive exploitation of large, sexually-mature fish.

Objectives of Regulations

There were four basic management objectives reported by the 11 states regulating paddlefish sport fisheries: 1) to regulate harvests; 2) to prevent overharvest; 3) to prevent excessive mortality; and 4) to protect brood stock.

The most frequently reported objective of paddlefish sport fishery management was to regulate harvests (eight states). Creel limits and seasons are the regulations most often imposed for this purpose. Preventing overharvest is an objective in six states, and again creel limits and seasons are employed. Six states attempt to reduce mortality by prohibiting high-grading, and one state protects brood stocks or populations by establishing closures and sanctuaries.

Two management objectives — prevention of overharvest and protection of brood stock — were reported by the seven states regulating paddlefish commercial fisheries. Four of these states prevent overharvest by using a variety of regulations including quotas, seasons, gear restrictions and closures. Two states attempt to protect paddlefish brood stock or populations with minimum size limits. One state uses regulations, but has not defined management objectives.

Of the 11 states regulating paddlefish sport fisheries and defining man-

Table 3. *Paddlefish sport fishing (snagging) seasons.*

State	Season	Angling days
Kansas	March 15-May 15	62
Missouri	March 15-April 30[1]	47
	March 15-May 15[2]	62
	October 1-December 31[2]	92
Montana	May 1-July 10	71
Nebraska (Above Gavins Point Dam)	November 1-April 30	182
(Below Gavins Point Dam)	November 1-March 31	152
North Dakota	April 30-July 1	63
South Dakota (Above Gavins Point Dam)	November 1-April 30	182
(Below Gavins Point Dam)	November 1-March 31	152

[1]Above Lake of the Ozarks
[2]Statewide

agement objectives, Kansas, Missouri and North Dakota indicate that effects of their regulations are known. Kansas and North Dakota report that using creel limits and seasons provides a regulated sport harvest, one of the objectives of their management programs. Missouri also reports that a combination of creel limits, seasons, closures, and sanctuaries fulfills their management objectives of protecting brood stock, preventing overharvest, and providing a regulated annual harvest of 3,500 to 4,000 fish (Graham[217]).

Other states generally concede that effects of sport fishing regulations on paddlefish populations are presently unknown. Effects of commercial fishing regulations on paddlefish populations are even more obscure. Only Kentucky reports that use of harvest seasons, closures and gear restrictions prevent commercial overexploitation.

Management Implications

Results of this survey show clearly that little is known about the effectiveness of regulations in protecting paddlefish. Although 14 states have placed regulations on sport or commercial harvest, only four indicate that their regulations are accomplishing management objectives. The other 10 report that the effects of their regulations are unknown.

With these limitations in mind, I offer the following evaluation of the effectiveness of current paddlefish regulations.

Sport Fish Regulations

Of the regulations imposed on paddlefish sport fisheries, only those that manage angler pressure, prohibit high-grading and establish closures and sanctuaries are likely to fulfill management objectives. High-grading regulations are effective in reducing snagging mortalities where this problem exists, because snagged fish cannot be released. Closures and sanctuaries are also effective if protection of a population, a portion of a population, or a spawning area is the management objective. Closures are sometimes imposed in areas where sport or commercial activities cannot be allowed because too little is known about the status of the paddlefish population.

Creel limits distribute harvest among anglers rather than reduce total harvest of a population (Everhart and Youngs 1981). Although the effectiveness of creel limits in regulating paddlefish harvest is unknown, this form of regulation has become more common as paddlefish populations have been subjected to increasing interest and pressure. Possession limits, however, are generally not effective and may merely reinforce creel limits (Fox 1975).

Sport fish seasons in present use are probably not effective either in regulating paddlefish harvests or protecting segments of the population. Most seasons are set to coincide with spring concentration and spawning, when the fish are most accessible to anglers. This limits their protective effect. However, seasons could be used to manage anglers, especially by alleviating conflicts between user groups.

Commercial Regulations

Commercial regulations which show the most potential for preventing overharvest and protecting brood stock are closures, gear restrictions and quotas.

Closures in commercial fisheries are effective in preventing harvest in areas where paddlefish concentrate and in reservoirs where the status of populations is unknown.

Maximum allowable mesh size is one of the more innovative gear restrictions used to prevent overharvest of sexually mature fish. Maximum mesh size allows continued commercial harvest of food fishes, including paddlefish, without adversely affecting paddlefish populations.

Quota systems can be effective in preventing overharvest, but sufficient data on a population is essential to establish realistic quotas. North Dakota is currently the only state using quotas for commercial fisheries. Their regulation allows incidental harvest of paddlefish, but prevents large-scale exploitation.

Minimum size limits may be one of the least effective commercial regulations, if the objectives are to limit harvest and protect brood stock. Minimum size limits currently in use are too low, because they allow fish to be harvested at 3-6 years of age, although sexual maturity is not reached until 7-12 years (Russell[446]). Brood stock may be better protected by maximum allowable mesh-size restrictions than by minimum size limits.

Effectiveness of seasons in commercial fisheries is still largely unknown. Only Kentucky reported that management objectives were met by the use of seasons. However, it seems likely that closed seasons during spawning periods would protect many adult fish from commercial exploitation.

Historically, commercial fisheries have been relatively unregulated, probably because of the low economic value of paddlefish. However, in the last few years the harvest of paddlefish has become profitable because of increased use of their roe for caviar. Commercial fishing pressure on populations has thus increased, and may result in additional regulations.

Conclusions and Recommendations

Of the regulations reviewed, creel limits are universally accepted for regulating angling pressure on paddlefish, although their effectiveness has not been evaluated. As paddlefish snagging continues to gain popularity, creel limits will remain a major management technique. In addition, little-used regulations such as harvest quotas show promise for preventing overexploitation, protecting brood stock, and regulating catch.

Harvest quotas have been well-developed in saltwater commercial fisheries and have been recommended for freshwater sport and commercial fisheries (Anderson 1975; Royce 1972). A quota system of harvest could be developed by adopting the 15% exploitation rate currently used to define overharvest (personal communication, 1983, L. Kim Graham, Fisheries Research Biologist, Missouri Department of Conservation, Columbia, Missouri). That portion of the population could be spread among anglers by current creel limit regulations. Creel surveys could be used to predict the time when the quota would be met.

Quota systems require the administrative flexibility to close a fishery at the time a harvest quota is met, regardless of the continued accessibility of the fish. They also require a thorough knowledge of the resource. When such knowledge is acquired, the quota can be adjusted by taking into consideration previous harvests and spawning conditions.

Gear restrictions, although presently used by only one state, show promise for managing commercial fisheries. Maximum mesh sizes, for instance, could be more widely used to prevent overharvest and protect brood stock (the two management objectives stated for commercial fisheries), while continuing to allow harvests.

This survey indicates how little is known about the effectiveness of current paddlefish regulations. It also identifies three ways to improve paddlefish management through regulation. These are:

1) Carry out more research on the effectiveness of current paddlefish regulations;
2) Develop more specific management objectives for paddlefish populations; and
3) Conduct a more thorough investigation of the status of paddlefish populations throughout their range.

References

ANDERSON, R. O. 1975. Optimum sustainable yield in inland recreational fisheries management. Pages 29-38 *in* P. M. Roedel, editor. Optimum sustainable yield as a concept in fisheries management. American Fisheries Society Publication No. 9, Bethesda, Maryland, USA. 39 pages.

EVERHART, W. H. AND W. D. YOUNGS. 1981. Principles of fishery science. Second edition. Cornell University Press, Ithaca, New York, USA. 349 pages.

FOX, A. C. 1975. Effects of traditional harvest regulations on bass populations and fishing. Pages 392-398 *in* H. Clepper, editor. National symposium on the biology and management of the centrachid basses. Sport Fishing Institute, Washington, D.C., USA. 533 pages.

MISSOURI DEPARTMENT OF CONSERVATION. 1951. Wildlife Code of Missouri — Rules and Regulations of the Conservation Commission. Revised to January 1, 1951. Missouri Department of Conservation, Jefferson City, Missouri, USA. 80 pages.

ROYCE, W. F. 1972. Introduction to the fishery sciences. Academic Press, New York, New York, USA. 351 pages.

The Culture of Paddlefish — A Review of Methods

L. Kim Graham, E. Jerry Hamilton[1], Thomas R. Russell
and Charles E. Hicks[2]

Missouri Department of Conservation
Fish and Wildlife Research Center
Columbia, Missouri 65201

Abstract

Procedures for rearing paddlefish (*Polyodon spathula*) were developed in Missouri during the mid-1960s and early 1970s. Since then, other fish culturists have made significant contributions or modifications to these methods, but our basic procedures are unchanged. The process involves: collecting adult paddlefish from the wild each spring; injecting females with hormones; collecting, fertilizing and hatching eggs; and rearing fingerlings in fertilized ponds, in polyculture with channel catfish (*Ictalurus punctatus*), or in troughs.

Because paddlefish have exacting spawning requirements, intraperitoneal injections of hormones are necessary to stimulate ovulation. We found that a single injection of two paddlefish pituitary glands from donors of comparable weight to the recipient was best. However, the initial results with a new hormone ($LH\text{-}RH_a$) appear promising.

Paddlefish do not ovulate all of their eggs at once. Since they release only a small quantity of eggs at a time, it may take 8-10 hours to complete ovulation after the first eggs are collected. Eggs are collected dry, fertilized, and treated with a Fuller's earth suspension to prevent clumping. Water-hardened eggs are hatched in plastic jars with the water flow adjusted to vigorously roll the eggs. Paddlefish eggs are sensitive to most chemicals, but fungus can usually be controlled by removing dead and infected eggs daily. Eggs hatch in 5-7 days at a water temperature of 65 F. Paddlefish fry must be transported and stocked in rearing ponds soon after they begin feeding (5-7 days after hatching) to ensure survival.

The best method for rearing larval paddlefish is to stock them in heavily fertilized ponds and harvest them later in the growing season as fingerlings. Ponds should be thoroughly dried before filling to discourage aquatic insects and aquatic vegetation. They should be routinely fertilized to encourage the establishment of zooplankton populations. Stocking rates can vary, but the best survival and growth is expected with rates of 20,000 paddlefish fry/a.

Paddlefish may also be reared in polyculture with channel catfish. This technique is better suited to southern catfish growers because ponds can be double-cropped without much additional effort and expense. Polyculture ponds do not require fertilization, because of the enrichment from excess catfish feed.

We did raise paddlefish in fiberglass troughs, where they were trained to eat artificial diets. Growth and survival were excellent, and disease was not a problem. A major disadvantage to tank rearing is that it requires a large amount of space to rear a few thousand fingerlings; and as the fish grow, they must periodically be thinned.

[1]Present Address: Blind Pony Hatchery, Rt. 2, Sweet Springs, Missouri 65351
[2]Present Address: P.O. Box 6847, Jefferson City, Missouri 65102

Culture of paddlefish (*Polyodon spathula*) was first attempted by Alexander[10] in the early 1900s, but his efforts were unsuccessful. It was not until 1961 that paddlefish eggs were collected, fertilized, and hatched (Purkett[410]). A year later, Meyer and Stevenson[329] attempted to induce ovulation of paddlefish using hormones, but they were unsuccessful. Needham[352] ovulated paddlefish and hatched some of the eggs in Missouri in 1963. Russell[444] continued Needham's work and developed techniques for inducing ovulation and rearing paddlefish. Those techniques are the basic procedures used today, although they have been modified and refined (Friberg[188]; Kallemeyn[278]; Brandt[63]; Russell[450]).

Although there has been considerable interest in producing paddlefish, few have attempted it because of the lack of brood fish and the uncertainty of success. South Dakota became interested in producing paddlefish in the early 1970s to maintain populations in reservoirs where recruitment was lacking. From 1972-1979 they produced a limited number of paddlefish using techniques developed in Missouri (Unkenholz[520]). In the late 1970s, the Tennessee Valley Authority (TVA) also reared a limited number of fingerling paddlefish (personal communication, 1983, Ronald Pasch, Biologist, Tennessee Valley Authority, Knoxville, Tennessee) as did Semmens[455] at Auburn University, Alabama. Besides the continuing efforts in Missouri, these are the only published accounts of attempts to culture paddlefish.

Artificial Propagation Techniques

Although artificial propagation techniques for paddlefish were originally developed over 20 years ago, few people are now raising this species. Paddlefish production is still more of an art than a scientific, predictable hatchery function. However, in the last 10 years interest has been renewed in the propagation of paddlefish for stocking in the wild and for caviar production. This report summarizes what is known about stimulating ovulation, collecting and hatching eggs, and producing fingerling paddlefish.

Collection of Brood Fish

Brood paddlefish are obtained yearly from wild populations because they are readily available and because those held in ponds do not develop eggs each year (Russell[450]). However, a private aquaculturist recently reported successfully ovulating pond-held paddlefish (personal communication, 1985, Ken Semmens, Fish Culturist, Auburn, Alabama). Being able to maintain brood fish in a pond would greatly aid paddlefish production. Perhaps these techniques can be worked out and implemented by those who have adequate ponds and food supplies available to sustain adult paddlefish year-round.

Brood fish do not have to be collected from natural reproducing and self-sustaining populations. They can be collected from paddlefish populations established and maintained by periodic stockings of fingerlings (Graham[218]). Because "wild" populations are generally found in public waters, regulations restricting the capture of these fish may constrain commercial producers.

Brood paddlefish should be obtained within 1-1½ months of their normal spawning time. They are easily captured in large-mesh gill nets (5-7 in. square mesh) set in deep water or other areas where paddlefish concentrate such as along the edges of river channels. Best success is when water temperatures are between 45-60 F. As water temperatures warm, nets should be checked every few hours to prevent mortality of large, gravid females. When conditions are not suitable for netting, the adults can be captured by snagging.

Handling and Holding

Paddlefish should be transported from the collection site to the holding facility in standard hatchery hauling units, large enough to accommodate several gravid females. Holding water should be chilled and aerated to maintain high oxygen levels. We add a 1% solution of sodium chloride and 10 ppm of Terramycin to the water to prevent disease and reduce stress. Gravid females should not be crowded, particularly if the hauling distance is substantial. As a general rule, we haul no more than 1 lb. of brood paddlefish per gallon of water and attempt to limit our hauling time to 4 hours or less.

At the hatchery, paddlefish are usually held in small ponds, ¼ a. or less in size, and are periodically treated with a fungicide. They are transferred from the ponds to indoor holding facilities for injections, so they can be easily observed and handled. Any tank large enough to permit the fish to swim continuously is suitable. Round metal tanks or plastic swimming pools 8 ft. in diameter and 2 ft. deep are ideal, and will accommodate three or four brood fish comfortably. Tanks should be kept inside the hatchery building and supplied with a continuous flow of water to maintain a constant temperature. We prefer a continuous flow of 60-65 F water.

Selection of Brood Fish

No external physical characteristics are totally reliable for determining the sex of paddlefish. Females are generally larger than males and have a distended abdomen in the spring. However, males in excess of 40 lb. can be easily mistaken for females. Males frequently have more tubercles on the head than females, but this characteristic is subjective and not a reliable criterion for determining sex.

Although internal visual examination of the gonads is the only totally reliable method for determining sex, we have found that the appearance of the urogenital opening is a good indicator. When the area is swollen, inflamed, and slightly flabby, it generally indicates a gravid female. We also probe the urogenital opening with the little finger. Easy insertion and presence of eggs usually indicate the female is mature. These two criteria are highly reliable.

In Missouri, female paddlefish generally mature at 10-13 years and weigh 40-60 lb.; males mature at 7-10 years and weigh 20-30 lb. We prefer females in the 40-lb. range because they are easier to handle than larger fish and they have 300,000-400,000 eggs. Gravid females in Alabama have been reported to mature at about 25 lb. (personal communication, 1985, Ken Semmens, Fish Culturist, Auburn, Alabama). These smaller females are ideal for handling, but contain fewer eggs than larger females.

The number of gravid females required for artificial propagation depends upon the number of eggs desired and success of ovulation. In Missouri, we generally collect 20-30 females and 10-15 males each spring. A large number of females is required because many do not ovulate, or release only a small quantity of viable eggs.

Hormone Injections and Ovulation

A variety of hormones have been

used to induce ovulation. We define ovulation as the release of sex products, including the release of milky fluid, ruptured or overripe eggs, and viable eggs. Hormones that have been tested or are presently being used include frozen whole paddlefish pituitary glands, homogenized paddlefish pituitary extract, dried carp (*Cyprinus carpio*) pituitary glands, human chorionic gonadotropin (HCG), luteinizing hormone (LH), follicle stimulating hormone (FSH), clomiphene citrate, and des-Gyl[10] (D-Ala[6]) LH-RH-ethylamide (commonly referred to as LH-RH_a).

Whole frozen paddlefish pituitary gland is the most widely used hormone source for stimulating ovulation (Russell[444]; Unkenholz[520]; Semmens[455]). Glands are easily collected from mature paddlefish caught by sport and commercial fishermen. The gland can be collected by using a core tool powered by a hand brace (Tsuyuki et al. 1964) to remove a 1-in. core from the cranium and expose the pituitary gland, or by using a hacksaw to remove a section of the cranium anterior to the braincase, thereby exposing the brain and attached pituitary gland. The latter method is preferred.

Pituitary glands are placed in individually numbered vials, quick-frozen with dry ice and stored at 0 F. Weights of donor fish are recorded so fish can be injected with glands from similar-sized donors. It is not necessary to segregate pituitary glands on the basis of sex of the donor. Frozen whole glands are thawed and injected intraperitoneally with 1.5 cc of deionized water using a #16 needle. The injection is made about 3 in. anterior to the symphysis of the pelvic fins and about 1 in. on either side of the mid-ventral line. It is not necessary to homogenize the glands because they are easily ruptured when forced through the needle.

Russell[444] initially used a dosage rate of one gland per injection and reported that the rate generally stimulated ovulation. Although most of the females receiving that dosage ovulated, most ovulated only a small portion of their eggs. Females injected with one pituitary gland per injection seldom released more than 1-3% (5,000-6,000) of their eggs. Russell later increased the dosage to two pituitary glands per injection and stimulated more complete ovulation. At this dosage rate, about 85% of the females released 5% (20,000) or more of their eggs. About 40% of the females released 25% (100,000) of their eggs, and 18% released 50% (200,000) of their eggs. One female released over 500,000 eggs. Generally, females that ovulated did so after one injection. Attempts to effect a more complete release of eggs through increased injections were unsuccessful.

The dose of two glands per injection has been used as the standard for inducing ovulation. We achieved a 68% ovulation rate (Table 1), and South Dakota (Table 2) and TVA (Table 3) had 64% and 30%, respectively, with this dosage. Fish in Missouri released more eggs than those in South Dakota or Tennessee (Tables 1-3).

Male paddlefish are injected to ensure release of viable sperm. Russell[444] reported that some males were spermiated by injecting carp pituitary glands, but paddlefish pituitary glands are more dependable. Injections of both one and two pituitary glands per injection were tested in Missouri. Males injected with two glands displayed an increased flow of milt and were productive longer than those injected with one gland. Milt could generally be stripped for 5-7 days, and after spermiation ceased, males could generally be re-stimulated by another injection. How-

Table 1. *Missouri's experience with artificial propagation of paddlefish using paddlefish pituitary injections, 1963-1978.*

Year	Number females injected	Number females ovulated	Viable eggs collected	Number of fry hatched	Percent of hatch	Number of fingerlings produced
1963	2	2	45,000	4,000	9	0
1964	2	0	0	0	0	0
1965	0	0	0	0	0	0
1966	7	5	15,000	0	0	0
1967	1	1	1,000	0	0	0
1968	10	7	275,000	1,000	<1	0
1969	12	5	515,000	60,000	12	0
1970	13	7	680,000	151,000	22	0
1971	10	7	680,000	182,800	27	0
1972	0	0	0	0	0	0
1973	14	10	2,127,800	443,700	21	37,700
1974	15	9	740,100	399,800	54	54,800
1975	13	5	941,600[1]	230,900	25	5,200
1976	17	11	1,314,900	462,600	35	2,100
1977	10	8	1,350,000[2]	513,500	38	3,100
1978[4]	42	36	7,400,000[3]	5,032,000	68	180,000
Total/ average	168	114	16,085,400	7,481,300	45	282,900

[1]One female was injected with carp pituitary and released 159,300 eggs. One additional female released 178,000 eggs and was not injected.
[2]Two females were injected with carp pituitary and released 203,000 eggs.
[3]Several females had only one developed ovary.
[4]After 1978, approximately 90 additional females were injected with two pituitary glands per injection and a high percentage of them ovulated viable eggs. From 1979-1985, approximately 1,254,000 fry and 330,700 fingerlings were produced.

ever, since one gland per injection was generally enough to stimulate adequate spermiation, our standard hatchery procedure is to inject all males with one pituitary gland. Those that don't spermiate after the first injection are not reinjected. Generally, six or seven males are sufficient to fertilize eggs collected from 10-20 females. Although most males weigh 16-40 lb., we prefer males in the 20-lb. range because they are easier to handle than larger fish.

Semmens reports that sperm can be collected and stored in a refrigerator (personal communication, 1985, Ken Semmens, Fish Culturist, Auburn, Alabama). He diluted it 1:200 with water immediately before fertilization. South Dakota and TVA generally collect fresh sperm each time eggs are to be stripped. We usually collect enough sperm to last for about 2 hours, and we have found that the rate of fertilization is influenced more by the condition of the eggs than by the condition of the sperm. When eggs are in good condition, we expect 80-85% fertilization.

A new and promising hormone, LH-RH_a, has recently been evaluated in Arkansas (personal communication, 1985, Ken Semmens, Fish Culturist,

Table 2. *South Dakota's experience with artificial propagation of paddlefish, 1972-1978.*[1]

Year	Number females injected	Number females ovulated	Hormone used	Eggs collected	Number of fry hatched	Percent of hatch	Number of fingerlings produced
1972	6	2	PP[2]	200,000	0	—	0
1973	6	5	PP	606,000	72,500	12	0
1974	10	8	PP	591,000	134,000	23	500
1975	13	7	PP, CP[3], CG[4]	638,500	23,000	4	0
1976	7	6	PP, CP	773,000	80,000	10	0
1977	5	3	PP	302,000	58,000	19	4,500
1978	3	1	PP	20,000	0	—	0
Total/ average	50	32		3,130,500	367,500	12	5,000

[1]Adapted from Unkenholz[520]
[2]Paddlefish pituitary (PP)
[3]Carp pituitary (CP)
[4]Chorionic gonadotropin (CG)

Auburn, Alabama) and by us in Missouri. LH-RH_a was first used for paddlefish in 1983 and has proven to be an excellent hormone for ovulating females. It is administered in two injections. The first is administered at a dosage of 0.01 mg/kg of body weight, and the second at 0.09 mg/kg, 12-15 hours later. Although LH-RH_a is relatively expensive ($19/mg), it appears superior to frozen paddlefish pituitary glands for stimulating ovulation. Our hatchery personnel, and Semmens, report that when LH-RH_a is injected into physiologically "ready" females, ovulation is usually complete. We haven't had this hormone long enough or injected enough females to determine its effects when injected into females that are not fully ripe. Overripe females ovulate a milky fluid or ruptured eggs when injected late in the season. Because of the manpower saved by not collecting pituitary glands, and the apparent effectiveness of LH-RH_a as an ovulating hormone, we began using it exclusively in 1984.

Carp pituitary glands have been used to ovulate paddlefish in Missouri (Russell[444]) and South Dakota (Unkenholz[520]). They have been used successfully to ovulate many species of fish (Clemens and Sneed 1962) and are commercially available. Russell injected females twice with dosages of carp pituitary ranging from 0.5-12.0 mg/lb. of recipient body weight, and 19 of 31 females ovulated. Most ovulating females released non-viable eggs. Best ovulation was achieved at doses of 2 and 3 mg/lb. of body weight. Seventeen of 23 females injected at these dosages ovulated, but only four fish released viable eggs. Egg numbers from these four fish ranged from 30,000-160,000 eggs. Females that were injected at the other dosage rates, and did not ovulate after two injections, were not injected a third time. Whether they were physiologically unready or the dosage rate was inadequate is not known. Unkenholz[520] used carp pituitary (2.0-6.5 mg/lb. of body weight) in 1975 and 1976 in South Dakota but discontinued using it because of poor results.

Table 3. *Tennessee Valley Authority's experience with artificial propagation of paddlefish using paddlefish pituitary injections, 1978-1980.*

Year	Number females injected	Number of females releasing more than 1,000 eggs	Number eggs collected	Number of fry hatched	Percent of hatch
1978	23	9	610,000	336,000	55
1979	26	7	250,000	190,000	76
1980	10	2	170,000	50,000	29
Total/ average	59	18	1,030,000	576,000	56

Other hormones, such as homogenized paddlefish pituitary extract, HCG, LH, and FSH were evaluated in Missouri from 1963-1971 for ovulating females but the results did not justify further study (Russell[444]). Clomiphene citrate was also evaluated in Missouri from 1973-1976 and was similarly discontinued because of poor results. Although homogenized paddlefish pituitary extract was used to successfully ovulate three of five female paddlefish (two at 12, and one each at 8, 16 and 20 cc dosage rates), only one released a significant number of eggs (100,000). The extract requires a considerable amount of time to extract and assay. HCG, used successfully to ovulate several fish species (Sneed and Clemens 1959; Stevens 1964), was evaluated as an ovulating hormone but none of the four females injected (two at 300, and one each at 500 and 1,100 IU dosage rates) ovulated. It was discontinued because of expense ($19.50/10,000 IU), and because it caused lesions on the sides of the fish. Unkenholz[520] reported similar results in South Dakota. LH and FSH were each injected into a female (2.2 mg/lb. dosage rates) and neither female ovulated viable eggs. Clomiphene citrate, which has been used to ovulate goldfish (*Carassius auratus*) (Pandey and Hoar 1972; Pandey et al. 1973), was injected into 10 females at dosage rates of 5, 10, and 12.5 mg/lb. Four of the fish ovulated; one after one injection, another after four injections, and two after five injections. All fish injected with clomiphene citrate died within 24 hours of the injections, except one female that lived 5 days.

Obtaining the Eggs and Milt

After brood fish are moved into the hatchery building and injected, they should be watched carefully to determine the beginning of ovulation. Females injected with two paddlefish pituitary glands usually begin to ovulate within 18-24 hours, but females injected early in the spring when eggs are not fully ripe may not ovulate for 30-36 hours. Females injected with LH-RH_a usually ovulate within 2-14 hours after the second injection, although some may not ovulate for 24 hours.

Males usually release sperm within 24 hours after injection and generally remain viable for at least 48 hours. Paddlefish sperm is watery and dilute compared to seminal fluids from most other fish. Hand stripping generally produces excessive volumes of sperm and usually renders the male unproductive for the remainder of the day. This problem can be alleviated by catheterization. This permits collection of only the amount of seminal fluid necessary to ensure fertilization and extends the productivity of males. Two or three ripe males should be available to ensure adequate fertilization.

Paddlefish do not ovulate all eggs in an ovary at one time. They ovulate a small portion of eggs periodically, and then release them. Consequently, complete ovulation, if it occurs, may take 8-10 hours or longer. Eggs collected during each ovulation should be fertilized immediately.

The procedure for collecting eggs involves several people. Ideally, two work with the female and a third holds the egg collecting pan and adds the sperm. The people working with the female stand inside the round tank, capture the fish by gently holding the rostrum and tail and slowly turn it belly-up. A hand over the vent prevents loss of eggs. When the female relaxes, she is gently cradled and raised out of the water with the head slightly higher than the tail. Her belly is gently stroked toward the vent and the eggs are collected in a plastic pan. If the female is ovulating well, the eggs will flow freely. The female is then released. Usually in 20-30 minutes additional eggs have been ovulated and the process is repeated.

After the eggs are collected, sperm is added and stirred for approximately 2 minutes to ensure adequate fertilization. A Fuller's earth suspension is then added to prevent clumping. The eggs are stirred slowly for another 15 minutes, and flushed with clean water to remove the excess silt. The eggs are then transferred to larger containers, allowed to water-harden about 1 hour, and transferred to hatching jars.

Incubation of the Eggs

Roberson incubators were initially used to hatch paddlefish eggs (Russell[444]) but plastic hatching jars are preferred (Russell[444]; Unkenholz[520]). The standard hatching procedure is to fill each jar with 2 qt. of eggs (100,000) and roll them vigorously using 2 gal. of water per minute. The rolling aids in hatching and helps reduce fungus. Fungus can be a serious problem on paddlefish eggs (Russell[444]; Unkenholz[520]; Brandt[63]), but can be controlled with treatments of formalin and a fungicide. We used formalin (1,667 ppm for 15 minutes in the flow-through method, in the early 1970s but discontinued it because of high numbers of malformed fry. Brandt[63] reported that hatching success of eggs and survival of fry were significantly greater for fungicide-treated eggs than for formalin-treated eggs. Now when eggs are treated, we use a fungicide. We have found that usually no chemical treatment is needed if the eggs are sufficiently treated with Fuller's earth to prevent clumping, if they are rolled vigorously in the hatching jars, and if the dead or infected eggs are removed daily.

Hatching time is directly affected by water temperature. Russell[444] reported that eggs hatched in 10-12 days when water temperatures were 52-58 F, and in 6-8 days when water was warmed to 60 F. We have found that the best temperature for incubating eggs is 65 F because at this temperature they will hatch in 5-7 days. This shorter hatching time also reduces fungus problems.

We arrange hatching jars so that fry are carried by the jar outflow into screened collecting tanks. Generally six jars empty into one tank. Air is added to prevent impingement of fry and egg shells on the screens and to ensure adequate oxygen levels in all corners of the tanks. Fry are held in these tanks until their mouthparts are well developed, peristalsis begins, and they are actively seeking food. These features become apparent after 7-9 days in 60 F water or in about 5-7 days at 65 F.

Transporting Larval Paddlefish

Paddlefish fry must be stocked as soon as possible after they begin to feed to ensure survival. Fry can be transported in oxygenated hatchery tanks if the trip is less than 6-8 hours. They can be shipped longer distances in plastic bags and styrofoam containers (at the rate of 5,000 fry and 2 ½ gal. of water per bag) if the total transportation time does not exceed 24 hours. Double plastic bags are recommended and should be sealed tightly with rubber bands after oxygen is added. Fry should be kept cool and protected from direct sunlight.

At the destination, plastic bags should be floated in the water to achieve thermal equilibrium. When the fry have been acclimated, they should be released immediately. Fry should be stocked on a cloudy day or after sundown to prevent sunlight shock. Released fry tend to remain near the surface and will be windrowed along the bank if there is wind. Because of these problems, we recommend stocking after sunset when the wind usually diminishes.

Rearing Young Paddlefish

Three rearing techniques have been used to produce paddlefish fingerlings. All involve stocking fry and rearing them until they reach a desirable stocking size. The first method involves stocking fry in ponds and fertilizing to maintain adequate zooplankton densities (extensive culture/pond method). The second method involves stocking paddlefish fry or fingerlings in the ponds with another species (polyculture), usually channel catfish (*Ictalurus punctatus*). The third method involves rearing fry indoors in troughs and tanks on artificial diets (intensive culture).

Paddlefish can be produced by any of these methods. All three techniques are reliable for rearing fingerlings. Each has advantages and disadvantages, depending upon hatchery facilities and objectives.

Extensive Culture

Pond preparation — Paddlefish fry can be stocked in an earthen pond of any size; however, we prefer ponds 1 a. or larger. We believe ponds 2-5 a. would be ideal because water quality, especially dissolved oxygen levels, should be more stable and aquatic vegetation less of a problem than in small ponds. We recommend that ponds be dried thoroughly (at least 10 days) before filling. Semmens[455] recommends spreading dry hay on the bottom as the pond dries to protect the pond from erosion.

Pond filling — Unless the water source is a well, the water should be filtered through saran screens of the appropriate mesh size to prevent additional introduction of other fishes. Ponds should be filled and fertilized 10-14 days before stocking to ensure an adequate zooplankton population. If the pond is filled too early, it can develop a large aquatic insect population. Filling ponds in stages should be considered. This is a technique whereby ponds are filled slowly (1/3 to 1/4 total volume each 20-30 days) as the fish population develops during the growing season. We have only limited information on incremental filling, but the concept appears to offer several advantages. Larval fish do not need an entire pond, initial fertilization is cheaper and the newly flooded surfaces promote zooplankton development.

However, heavy rains can erode exposed pond banks while filling so they should be mulched for protection.

Pond fertilization — Fish production in ponds depends largely upon water fertility and water quality. Because hatchery ponds have no watershed, they must be fertilized to provide necessary nutrients for development and maintenance of dense zooplankton populations. The food of paddlefish fry and fingerlings is almost exclusively zooplankton and small aquatic insects (Russell[446]). The major zooplankton groups found in production ponds are cladocerans, copepods, and rotifers. Cladocerans (primarily *Daphnia*) are the preferred food of paddlefish (Michaletz[332]), probably because they are relatively large, are slow swimmers and are more easily captured. When paddlefish are about 5 in. TL, they begin filter-feeding and are able to utilize the faster swimming copepods. Both cladocerans and copepods have high reproductive rates and respond to fertilization.

Many different fertilizers and rates of application have been used to enhance the production of food organisms in hatchery ponds. Because the kinds and rates vary depending upon geographic area, information should be obtained from local aquaculturists or your University Extension Center.

Organic fertilizers seem most reliable for developing and maintaining cladocerans. Various hays, alfalfa meal or pellets, soybean and cottonseed meals, rice and wheat bran, brewer's yeast, and chicken, sheep and cow manure have been used individually and in combination to fertilize hatchery ponds. Application rates are best determined for each situation, but in general most organic fertilizers are applied at rates between 50-500 lb./a. These fertilizers can be applied at one time or at weekly intervals during the growing season. We prefer weekly treatments so we can monitor water quality and the development of zooplankton and adjust fertilizer applications accordingly.

Inorganic fertilizers (N-P-K), usually available in granular, powder, or liquid forms, have been used in combination with organic fertilizers but their use should be regulated carefully because they tend to promote rapid phytoplankton blooms. If organic fertilizers are used, the pond water should be carefully monitored for high pH and extremely high and low dissolved oxygen levels; both are detrimental to young paddlefish.

In Missouri, we fertilize when we fill the pond, about 10-14 days before the paddlefish fry are stocked, and then weekly with 150 lb./a. each of dehydrated alfalfa meal and dehydrated cow manure, 50 lb./a. of brewer's yeast, and 25 lb./a. of fresh chicken manure throughout the growing season.

We have never had problems establishing zooplankton by early filling and fertilization. However, Semmens[455] increased the initial numbers of zooplankton in production ponds in Alabama by culturing zooplankton and stocking 6-14 cladocerans/l into the production pond. This practice appears to have merit, especially where zooplankton are difficult to establish. He reported best success at establishing cladoceran populations by fertilizing weekly with 200 lb./a. of rice bran and 100 lb./a. of fresh chicken manure. He adjusted his weekly fertilization rates based on zooplankton development, water color, and dissolved oxygen concentrations, and reported that ponds with high zooplankton populations usually were a brownish-green color. Color may not be a valid criterion everywhere. We have found that some ponds with a brownish-green color have high phytoplankton and low zooplankton populations. Semmens also stopped fertilizing when dissolved oxygen concentrations fell below 3 ppm. We usually stop fertilizing when mid-day dissolved oxygen concentrations approach saturation (10-20 ppm) or when early morning dissolved oxygen concentrations drop below 4 ppm.

Pond stocking and growth — We stock production ponds with larval paddlefish when they begin to actively seek food, usually about 5 days old. We have evaluated stocking 10,000, 20,000, and 30,000 fry/a. and found that survival from fry to advanced fingerling size (10-12 in. TL) is best at the 20,000/a. rate. Annual survival was variable, but averaged 20-50%. Semmens[455] also reported high variability in survival of young paddlefish (0-50%) and attributed it to water quality. Since 1980, we have stocked all production ponds with 20,000 fry/a. and have averaged about 35% survival. Growth of young paddlefish is rapid. When zooplankton populations are high, they can grow 2.4 mm (0.1 in.)/day (Michaletz[332]) and at the end of the 140-day growing season, most of the paddlefish range from 10-12 in. TL. These growth rates are similar to those reported in Alabama by Semmens[455] and in South Dakota by Unkenholz[520].

Harvesting paddlefish fingerlings — Paddlefish fingerlings are easily harvested by seining or draining the pond. Because they are sensitive to poor water quality, they should not be handled when water temperatures exceed 70 F or be allowed to remain in muddy water for very long. Semmens and Shelton[456] recommended that paddlefish be harvested by seining because they are docile and easily captured. Paddlefish raised with channel catfish can be removed first by using a large-mesh seine that retains the paddlefish and lets the catfish escape.

We generally harvest paddlefish fingerlings by draining. We prefer to harvest fish in early morning or late evening when the air is cool and the water temperature is 60-70 F. Ponds should be drained slowly to prevent paddlefish from becoming impinged on the draining screens. Ponds with draining kettles that have cement floors are ideal because fish can be confined and fresh water can be added.

Paddlefish must be graded by hand. They are weak swimmers, stress very easily, and will not swim through conventional fish graders. After the fingerlings have been harvested and graded, they should be stocked within 48 hours because they do not hold well without feed and mortality increases rapidly. Fingerlings are transported in well-aerated hauling units, in a solution of 0.5% NaCl and 10 ppm Terramycin.

Potential pond problems — Rearing paddlefish in ponds is never as simple as stocking fry in the spring and harvesting them as fingerlings in the fall. Production ponds must be observed daily to monitor growth and survival and to detect potential problems. Predation by various vertebrates and growth of aquatic vegetation are easily recognized, but water quality and disease problems require more thorough observation. These problems can result in a total loss of production if not recognized early and corrected immediately. Good management is the key to successful production of paddlefish fingerlings in ponds.

• *Aquatic insects* — Predacious aquatic insects such as water boatmen, backswimmers, whirligig beetles, and (in some cases) water scorpions will eat paddlefish fry. Many of these insects can be controlled by drying the pond thoroughly before filling and by not refilling until 10-14 days before paddlefish fry are stocked.

We use this drying and filling regime and seldom have aquatic insect problems. When a problem occurs, a 50% mixture of diesel fuel and used motor oil applied at a rate of 5 gal./a. will quickly eliminate the insects and not harm the young paddlefish. We bubble air through the mixture for at least 24 hours prior to application to speed the breakdown of hydrocarbons which are toxic to paddlefish fry and zooplankton.

Commercial insecticides will control aquatic insects, but they should be used cautiously because most of them are toxic to zooplankton. Methyl parathion (0.5 ppm) or Baytex (2.5 ppb) will effectively control aquatic insects but they should be used cautiously because they can also cause deformities of fish if dosage rates are exceeded. These chemicals should be applied before the pond is filled or before the fry are stocked. If they are used after fry are stocked, they should not be applied until fry are 20-30 days old.

• *Clam shrimp* — These crustaceans can be a serious problem in paddlefish production ponds. Populations can become extremely dense about the same time in the spring as high zooplankton densities are needed for young paddlefish. Dense infestations of clam shrimp can cause the water to become muddy because of their burrowing in the pond bottom, and they will eat zooplankton. Baytex applied at a rate of 2.5 ppb has been used to control clam shrimp. This application rate will kill clam shrimp within 48 hours. However, because of its extreme toxicity to zooplankton, Baytex must be used cautiously.

• *Vertebrate predators* — Some birds are serious predators on small paddlefish. When a problem develops it must be controlled or significant numbers of paddlefish will be eaten. Herring gulls and black terns are major predators on paddlefish less than 2-3 in. TL. Belted kingfishers, green-backed herons, and great blue herons consume larger paddlefish (3-6 in. TL). These and other birds can be discouraged with netting stretched over the pond or with propane scare cannons. We also routinely control turtles, bullfrogs, and water snakes because they all eat small paddlefish. Muskrats should be eliminated from paddlefish production ponds because they not only eat small paddlefish, but also attract mink, which are efficient predators.

• *Aquatic vegetation* — Aquatic macrophytes and filamentous algae can be a serious problem in paddlefish production ponds. Young paddlefish are weak swimmers and easily become entangled and die. Several commercial herbicides can be used, but only in accordance with the manufacturer's label instructions. Most aquatic herbicides are toxic to zooplankton and should be applied before paddlefish fry are stocked or after the fry are 20-30 days old.

We treat paddlefish production ponds the day of filling with 1 ppm Aquazine. This one-time treatment is effective in controlling filamentous algae during the entire fish growing season.

• *Water quality* — Good water quality must be maintained in paddlefish production ponds, especially if the ponds are fertilized. Dissolved oxygen should be maintained above 4 ppm. If it drops below this, fertilization should be discontinued. Oxygen levels can be increased by adding fresh, aerated water or by mechanical aeration devices. If an aeration device is used, it should be screened to prevent young paddlefish from becoming injured by the impellers. Some hatcheries are equipped to inject air at the pond bottom to alleviate low dissolved oxygen problems without harming the paddlefish.

Other chemical parameters such as alkalinity and total hardness have never been a problem in Missouri. Piper et al. (1982) reported that fish grow well over a wide range of alkalinity and hardness, but values of 120-400 ppm are optimum. They can be modified, however, by adding agricultural lime to increase the buffering capacity of the water and to neutralize acid soil-water interactions. Water temperature in production ponds will vary considerably with location; however, optimum temperatures for paddlefish growth are between 65-90 F. Temperatures above 90 F frequently cause zooplankton numbers to crash which affects growth of paddlefish. High temperatures can sometimes be lowered by adding cooler water.

- *Diseases* — Diseases are generally not a serious problem. We have had occasional minor problems with external parasites, primarily *Trichodina, Scyphidia* and *Costia.* These parasites can be effectively treated with potassium permanganate ($KMNO_4$) at a 2 ppm dosage rate until enough chemical has been added to maintain a light brown color. Copper sulfate ($CuSO_4$) is not recommended for treatment of diseases in young paddlefish because it is extremely toxic. Formaldehyde should not be used at rates above 75 ppm because it will stress the fish.

Polyculture

Polyculture is a techique in which two or more non-competing species of fish are raised in the same pond, allowing efficient use of natural or prepared feeds. Polyculture has long been practiced in Asia and various South American countries where annual production must be maximized, but it is a relatively recent concept for fish culturists in the United States. Aquaculture in the United States is currently monoculture-oriented (Shelton and Smitherman[461]), however, aquaculturists in several southern states have successfully reared channel catfish with several species of Chinese carp. Channel catfish are popular for pond culture because they readily eat commercial diets and appear to be ideally suited as a polyculture species. Ponds stocked with both channel catfish and paddlefish do not require fertilization because excess catfish food stimulates zooplankton production (Kirkendahl[283]). Since paddlefish are not attracted to pelleted catfish food, these two species do not compete.

We reared paddlefish with channel catfish from 1973-1975. Ponds were stocked at various rates and with fry and fingerlings. Paddlefish fry were stocked at 7,000 and 10,000/a. with catfish fry at 10,000/a. Paddlefish fingerlings (3.5-6.0 in. TL) were stocked at 2,000, 3,000, and 4,000/a. with 50,000, 70,000, and 100,000 catfish fry/a., respectively. Additionally, paddlefish fingerlings (3.5-6.0 in. TL) were stocked at 2,000, 3,000, and 4,000/a. with 1,000 catfish fingerlings (4.0-6.0 in. TL).

We were not able to determine which stocking rates were best because of extreme variation in paddlefish survival (2-77%). Paddlefish from ponds with lowest survival (2%) averaged 19 in. TL when harvested after a summer's growth (about 140 days), whereas in ponds with best survival (77%) paddlefish averaged only 12 in. TL. Paddlefish fingerlings stocked at the lowest densities (2,000/a.) were no larger than paddlefish stocked as fry (7,000 and 10,000/a.). At the end of one season's growth both groups of young paddlefish averaged about 12 in. TL. Channel catfish grew well in all ponds, had excellent survival, and did not appear to be affected by the paddlefish.

Currently, we are not rearing paddlefish with channel catfish although the technique has potential. The Missouri Department of Conservation is committed to producing 30,000 advanced fingerling paddlefish (10-12 in. TL) each year for stocking in public waters to maintain sport fisheries. Because of limited hatchery space and the large number of paddlefish required each year, we will continue to use the monoculture technique to rear more paddlefish in fewer ponds.

Intensive Culture

Intensive culture in the United States was accelerated in the 1960s when fish culturists discovered that certain fish species could be confined in tanks or troughs and reared on artificial diets rather than natural foods. Intensive culture has at least two important advantages over pond culture: investments in ponds and feeds can be reduced, and the environment can be controlled. Coolwater species such as walleye (*Stizostedion vitreum*) and tiger muskie (*Esox masquinongy* × *E. lucius*) are receiving most of the attention in intensive culture, but other species such as channel catfish are routinely reared in tanks on artificial diets. The development of new artificial diets has stimulated interest in intensive fish culture. Culturists now can use numerous diet formulations and sizes, allowing them to rear fish species that previously would not eat commercial diets.

There are many advantages in using formulated feeds. They are available throughout the year; their application is easily regulated and can be automated; they are convenient to store, handle and augment with medications; they do not introduce parasites and diseases; and they generally reduce the cost of the feeding program (Orme 1978).

Purkett[410] was the first to feed paddlefish an artificial diet. He fed Purina trout starter to fry, but concluded the diet was not adequate because of high mortalities. Russell[444] fed an Abernathy salmon diet to paddlefish fry with similar results. Fry were attracted to the feed and fed actively but there was a continual loss of fish and eventually almost all died. Russell suggested that paddlefish could be fed a commercial diet if one were developed containing adequate vitamins, minerals, and amino acid supplements. In addition, he suggested developing a soft diet because some paddlefish ingested the hard pellets, but then spit them out. Friberg[188] and Kallemeyn[278] experienced similar results feeding artificial diets to paddlefish fry in South Dakota.

We reared paddlefish in tanks and fed them artificial diets in 1975 and 1976. Fry were stocked at the rate of 1,000, 2,000, and 4,000 in rectangular fiberglass troughs (10 ft. long, 2.5 ft. wide, and 12 in. deep) holding 125 gal. water. Because data for both years are similar, only results for 1976 are summarized and presented (Table 4). Water temperatures during the growing season ranged from 63 F in the spring to near 85 F in mid-summer. Troughs were cleaned daily and hatchery lights were always on.

We began feeding the paddlefish fry immediately after they were stocked in the troughs. Zooplankton, mostly *Daphnia,* were seined from a nursery pond to provide the initial food (Table 4). *Daphnia* were maintained at high levels in the troughs during the first week to ensure an adequate food supply at all times. Once fry were actively eating zooplankton, we added an experimental cool-water sinking diet (W-7), developed at the U.S. Fish and Wildlife Service Diet Development Center in Spearfish, South Dakota. Within 3 days, some paddlefish fry were eating the prepared diet (No. 1 fry starter, 3/32 in. or 2.5 mm pellets), and within 13-18 days most were accepting the diet.

Artificial feed was dispersed by automatic feeders every 30 minutes and by occasional handfeeding. Brandt[63] fed paddlefish with automatic feeders programmed to release feed each hour, but suggested feedings should be more often for better growth and survival. Feed sizes were increased when it became apparent the paddlefish could accept a larger size.

Young paddlefish ate the W-7 diet when it was dropped close to them but they would not search for it. Crowding screens were used to concentrate fish under the automatic feeders. As food dropped, the fry swam excitedly through the water. They did not display the typical feeding activity exhibited by channel catfish or other species trained to eat artificial diets. Instead, they turned slowly on their sides or backs as they randomly selected the diet. We observed that some paddlefish ate the diet on the surface before it sank, some fed in midwater as the pellets sank near them, and others actually fed near the bottom of the trough. Brandt[63] noted similar feeding habits of paddlefish fry in South Dakota. He reported that the artificial feed was only available as it sank through the water column and was not eaten once it reached the bottom. He suggested that diets be developed that would remain in sus-

Table 4. *Survival, growth and diet of paddlefish reared in rectangular troughs and later in round tanks, Missouri Department of Conservation, Blind Pony Hatchery, Sweet Springs, Missouri, April-July, 1976.*

Day	No. Fish	Survival (%)	TL (in.)	Diet
RECTANGULAR TROUGH PRODUCTION				
1	4,000	100	<1.0	Zooplankton
7				Zooplankton and W-7, #1[1]
16				Zooplankton and W-7, #2
24	2,337	58	2.9	Zooplankton and W-7, #2
31	1,965[2]	49	4.6	Zooplankton and W-7, #2
32				Zooplankton and W-7, #2
45				W-7, #3
48	1,775	44	5.2	W-7, #3
ROUND TANK PRODUCTION				
49	1,775	44	5.2	W-7, #3
65				W-7, #4
82				W-7, 3/32 in.
88	1,692[3]	42	10.3	W-7, 3/32 in.

[1]Experimental coolwater sinking diet developed at the U.S. Fish and Wildlife Service Diet Development Center, Spearfish, South Dakota. Numbers 1, 2, 3, 4, and 3/32 in. denote feed sizes.
[2]*Costia* outbreak on day 26 and 27 caused high mortality.
[3]At the end of the 88 day growth period, the total weight of surviving paddlefish was 195.2 lb.

pension, and be available to the fry for a longer period of time.

Once the fry were eating the W-7 diet, zooplankton was added in decreasing amounts twice daily for about 30 days, then once daily for an additional 10 days, then discontinued (Table 4). During both years (1975-1976) we fed zooplankton for a total of 35-42 days before switching the fish exclusively to artificial feed. During both years, survival after 30 days was about 50%. At this time the fingerlings were about 4.5 in. TL.

Crowding of small (2-3 in. TL) paddlefish in troughs can result in two unusual behaviors; billing and tailbiting. Both are symptoms of overcrowding. Billing is characterized by small paddlefish swimming at the surface with their rostrums out of water. When paddlefish begin billing, they have quit feeding and will soon die unless they are thinned. Tailbiting was also observed when paddlefish were crowded. Brandt[63] also reported tailbiting but suggested it was attempted cannibalism. However, we have found that when small paddlefish were transferred to larger tanks both billing and tailbiting were no longer a problem.

We had some problems with small paddlefish concentrating in the corners of the rectangular troughs. This problem can lead to crowding and will result in reduced feeding and slow growth rate. We believe crowding in corners can be alleviated by tank design. As paddlefish grew and the density per trough increased, they were periodically hand-graded by size and transferred to other troughs. Disease was seldom a problem in troughs.

Once the young paddlefish grew to about 5 in. TL, they were transferred to round production tanks (5 ft. diameter, holding 250 gal. water) for the remainder of the growing season (Table 4.) Round tanks alleviated crowding because of design, and mortality was insignificant. Fine mesh screens were placed over the tanks to prevent paddlefish from jumping out.

At the end of the growing season (1975 and 1976), survival was 34 and 42%, respectively. During 1976 growth of over 0.1 in. per day was considered excellent (Table 4). The only growth lag we experienced was during 1975 when fish did not grow appreciably for several days after we switched them from zooplankton to the artificial diet. We did not observe a growth lag in 1976. At the end of both growing seasons, all fish exceeded 10 in. TL. The 1,692 reared in 1976 (88 days, Table 4) averaged 2 oz. each.

Intensive culture of paddlefish is relatively new and requires careful attention. Paddlefish can be trained to accept artificial diets, but cannot be crowded. They must be graded and sorted frequently, and require a relatively large amount of space. An entire hatchery can easily be tied up producing only a few thousand fingerlings. Considerable manpower is required daily to feed fish and clean tanks. Although artificial feed is rather inexpensive, total costs are probably higher than purchasing fertilizer for pond culture. One major advantage to intensive tank culture is that all fish can be observed. At our highest stocking rates (4,000 fry/trough), we reared over 30 fish per gal. of water (tank capacity). Crowding potentially encourages diseases, but careful daily monitoring usually prevents excessive mortality.

Although we currently rear large numbers of paddlefish fingerlings each year using the monoculture, pond-method, we would prefer to raise them intensively in tanks. Having the fish inside a hatchery building under direct observation allows hatchery personnel to check the fish often and potential problems can often be alleviated before mortality occurs. Hatchery space is presently our limiting factor.

Conclusions

Procedures for rearing paddlefish were developed in Missouri about 15-20 years ago and only slight modifications have been made since then. The rearing process involves collecting wild brood stock each spring, injecting females with hormones, collecting, fertilizing, and hatching eggs, and rearing fingerlings. Young paddlefish can be reared in fertilized ponds, in polyculture with channel catfish, or in troughs and tanks on an artificial diet.

We presently rear paddlefish in earthen ponds fertilized to encourage the establishment of *Daphnia*. We have found that our best survival and growth (20-50% survival resulting in 10-12 in. TL fingerlings after 140 days) is achieved by stocking 20,000 fry/a. in ponds 1 a. or larger. We are currently not rearing paddlefish in polyculture with channel catfish because we can rear more paddlefish in fewer ponds using the monoculture method. We have successfully reared fingerlings intensively in troughs and tanks but do not at present because it requires a tremendous amount of hatchery space. If space were not a problem we would prefer to rear them intensively, because having the fish inside a hatchery building permits close observation by hatchery personnel.

References

CLEMENS, H. P., AND K. E. SNEED. 1962. Bioassay and use of pituitary materials to spawn warmwater fishes. United States Fish and Wildlife Research Report 61. 30 pages.

ORME, L. E. 1978. The status of coolwater fish diets. Pages 167-171 *in* R. L. Kendall, editor. Selected coolwater fishes of North America. American Fisheries Society Special Publication, Number 11. Bethesda, Maryland, USA. 437 pages.

PANDEY, S. AND W. S. HOAR. 1972. Induction of ovulation in goldfish by clomiphene citrate. Canadian Journal of Zoology 50(12):1679-1680.

PANDEY, S., N. STACEY, AND W. S. HOAR. 1973. Mode of action of clomiphene citrate in inducing ovulation of goldfish. Canadian Journal of Zoology 51(12):1315-1316.

PIPER, R. G., I. B. MCELWAIN, L. E. ORME, J. P. MCCRAREN, L. G. FOWLER, AND J. R. LEONARD. 1982. Fish hatchery management. United States Fish and Wildlife Service, Washington, D.C., USA. 517 pages.

SNEED, K. E., AND H. P. CLEMENS. 1959. The use of human chorionic gonadotropin to spawn warmwater fishes. The Progressive Fish-Culturist 21(3):117-120.

STEVENS, R. E. 1964. A final report on the use of hormones to ovulate striped bass, *Roccus saxitilis* (Walbaum). Proceedings of the Annual Conference Southeastern Association of Game and Fish Commissioners 18:525-538.

TSUYUKI, H., P. J. SCHMIDT, AND M. SMITH. 1964. A convenient technique for obtaining pituitary glands from fish. Journal of the Fisheries Research Board of Canada 21(3):635-637.

Establishing and Maintaining Paddlefish Populations by Stocking

L. Kim Graham

Fish and Wildlife Research Center
Missouri Department of Conservation
Columbia, Missouri 65201

Abstract

The native range of the paddlefish (*Polyodon spathula*) has decreased because of destruction of spawning areas, blockage of movements by dams, and in some cases, pollution and overexploitation. Some major paddlefish populations have suffered irretrievable losses due to these influences.

Several midwestern states have considered stocking fingerlings to reestablish paddlefish sport fisheries, but few have a dependable supply of hatchery-produced fingerlings. At this time, the only documented stocking effort has been the successful introduction of paddlefish into Table Rock Lake, Missouri.

Paddlefish fry and fingerlings were stocked in Table Rock Lake (43,100 a.) in 1970 and 1972-1977, respectively, to determine if a population could be established. This was the first major attempt by any agency to introduce paddlefish into previously unoccupied waters. An excellent population has developed, growth is good, and several fertilized eggs were found during 1983. A sport fishery is also developing and becoming more popular each year. During the springs of 1983 and 1984, approximately 2,000 and 970 paddlefish, respectively, were caught by anglers as the fish concentrated in the upper portions of the James River Arm. Thus, it appears that populations of paddlefish can be established in new waters or maintained in environmentally impacted waters by maintenance stockings.

Stocking of adult paddlefish would not be economically practical because availability of large fish is limited. Fry should be stocked only in waters where there is an abundance of food for them, and where predator fish populations are low. In Missouri, the best stocking results are achieved with fingerlings 10-12 in. total length (TL) at densities of two to four fingerlings per acre. Although paddlefish can be stocked into smaller ponds and lakes as a novelty with limited economic return, they are best suited for larger lakes where sport fisheries will develop, or where they can be commercially harvested for flesh and roe. If possible, lakes should have tributary streams with spawning substrate for natural reproduction.

Present demands on the resource indicate that additional paddlefish populations for both sport and commercial use should be developed. With dedication, commitment and proper management techniques, resource managers can ensure that sport and commercial paddlefish populations will prosper.

Since the turn of the century, the range of paddlefish (*Polyodon spathula*) has diminished (Gengerke[195]) because of destruction of spawning areas, blockage of movements by dams, and in some cases, pollution and overexploitation (Pflieger[398]; Carlson and Bonislawsky[95]). In many instances these changes have caused irretrievable losses to major paddlefish populations. Once common in large rivers of the Mississippi River Basin, paddlefish are now found in large numbers only where they reproduce naturally.

Because of increased public awareness of these threats to paddlefish habitats, and a reevaluation of fishery management priorities by several fish and wildlife agencies, there has been an increased interest in paddlefish. Although not all states within the native range of the paddlefish consider this species a resource worthy of management, those states with continuing or renewed interest in managing their stocks of paddlefish are advocating stricter harvest regulations and habitat protection. In Missouri, paddlefish populations are being reestablished by stocking.

The closure of Harry S. Truman Dam in 1977, on the Osage River in Missouri, inundated all known paddlefish spawning areas in the Osage River Basin. The dam also trapped most of the paddlefish downstream in Lake of the Ozarks, where there are no spawning areas (Russell[450]; Graham[217]). An excellent tailwater fishery developed below Truman Dam, but without natural reproduction this fishery will gradually disappear. A similar situation occurred in South Dakota, where excellent paddlefish fisheries developed in the tailwaters of three Missouri River main-stem reservoirs after closure. These fisheries have since declined drastically because natural spawning areas were inundated (Unkenholz[520]).

It is doubtful whether a paddlefish population will develop in Truman Reservoir. There are few adults in the reservoir, and potential spawning substrate in the upper Osage River and other smaller tributaries is extremely limited. Because of the documented problems in South Dakota and Missouri, future sport fisheries for paddlefish in the Osage River Basin will depend upon hatchery stockings.

To determine whether a paddlefish population could be established and maintained in the Osage River Basin through stocking, fingerling paddlefish were stocked into Table Rock Lake. This was the first major attempt by any agency to introduce paddlefish into previously unoccupied waters.

Stocking Strategies

Table Rock Lake, a 43,100-a., relatively clear Ozark reservoir located in southwestern Missouri (Fig. 1), was selected for this study because paddlefish were not present in the lake. The 12,000-a. James River Arm contains large areas of chert gravel that appear suitable for spawning, and the watershed is extremely fertile. The drainage area of Table Rock Lake is composed of Ozark mountains, prairie and domestic grasslands, tilled farmlands, and Missouri's third-largest metropolitan area, Springfield. Streams that feed Table Rock Lake are characteristically clear, gravel-bottomed and spring-fed. Flash floods commonly follow heavy precipitation. The James River is the most eutrophic of the tributary systems (Dieffenbach and Ryck 1976) and supports a dense zooplankton population.

Paddlefish fry (6-8 mm, 0.2-0.3 in. total length — TL) were stocked into the James River Arm of Table Rock Lake in 1970, followed by annual

Figure 1. *Table Rock Lake, Missouri.*

stockings of fingerling paddlefish (10-12 in. TL) from 1972-1977 (Table 1). The largest releases were made in 1973 and 1974, when 25,000 and 38,000 fingerlings, respectively, were stocked. During the 6-year period nearly 83,000 fingerling paddlefish were stocked, representing an initial stock density of about two fish per acre.

Population Development

Monitoring Methods

Fishermen reported taking or seeing paddlefish incidentally before we began our sampling program during the spring of 1978. Because of these early reports, we suspected that survival and growth were good. We sampled Table Rock Lake each spring from 1978 through 1984 using 6 and 7-in. bar mesh sinking gill nets. Greatest effort was expended in the upper James River Arm where most of the fish were concentrated, but paddlefish were captured in all arms of the reservoir. All paddlefish were weighed and measured using body length (anterior orbit of eye to fork of tail, Ruelle and Hudson[432]), jaw-tagged with numbered Monel metal bands (#16), and released at the site of capture. Dentary bones were collected from dead fish and from fish caught by cooperating fishermen for aging purposes (Adams[5]). During the 7-year sampling period (143 net-days), 679 paddlefish ranging from 7-70 lb. (Fig. 2) were captured.

Table 1. *Numbers of fingerling paddlefish stocked in Table Rock Lake, Missouri, 1972-1977.*[1]

Year	Number stocked
1972	10,000
1973	25,000
1974	38,000
1975	5,200
1976	1,900
1977	2,500
TOTAL	82,600

[1]In 1970, 100,000 paddlefish fry were stocked but survival was apparently poor.

Harvest data were also used to monitor population development. Statewide fishing regulations permitted the harvest of paddlefish from March 15–May 15 and from October 1-December 30 each year, but sport snagging did not become popular until 1982, when the fish grew to an acceptable size and stimulated interest. Prior to 1982, most of the paddlefish were small and caught only incidentally on trotlines or snagged accidentally while fishing for other species.

Survival

It appears that few, if any, paddlefish fry from the 1970 stocking survived. We have never collected fish from that year-class. Since paddlefish fry are pelagic, we believe that they were eaten by the abundant population of predacious fishes in Table Rock Lake.

Survival of stocked fingerlings (10-12 in. TL), on the other hand, appears to have been high. Each year following releases, small paddlefish were observed by many anglers, and several fingerlings were caught by crappie fishermen using minnows. Netting also revealed that survival of stocked fingerlings was high, although we did not attempt a population estimate. During our net sampling from 1978-1984, we captured and tagged 679 paddlefish. Only four of these fish were recaptured. The lack of spawning in this developing population is becoming apparent, because there are few fish smaller than 30 lb. (Fig. 2).

Growth

Paddlefish growth in Table Rock Lake has been excellent. Average weight of paddlefish increased yearly from 1978 through 1984 (Fig. 2) because of rapid growth, an increase in numbers of females reaching sexual maturity, and lack of natural annual recruitment. A 7-lb. increase (29 to 36 lb.) in average weight from 1980 to 1981 was caused in part by increased numbers of gravid females. The percentage of females in the net catch increased from 14% in 1980 to 37% in 1984.

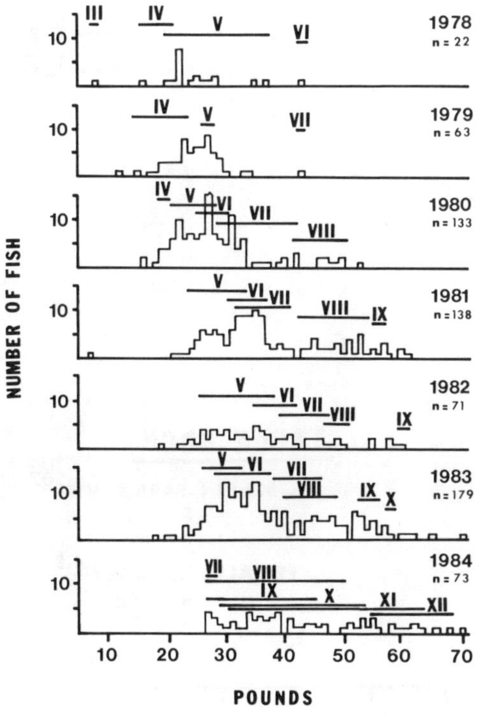

Figure 2. *Weight frequency and age range of paddlefish captured in gill nets, Table Rock Lake, Missouri, 1978-1984.*

Growth of paddlefish in Table Rock Lake was faster than that reported from many other areas, including the Missouri River in South Dakota (Sprague[481]; Unkenholz[517]), the Yellowstone River in Montana (Rehwinkle[419]), the Mississippi River in Iowa (Gengerke[195]), the Neosho River in Kansas (Bonislawsky[55]), Grand Lake of the Cherokees in Oklahoma (Combs[112]), Old Hickory Reservoir in Tennessee (Pasch[392]), and the Osage River in Missouri (Purkett[409]). This relatively rapid growth was probably enhanced by the long growing seasons and an abundant food supply.

Age-weight relationships for paddlefish collected from Table Rock Lake and Lake of the Ozarks indicate that both males and females from Table Rock Lake were heavier for a given age than were those from Lake of the Ozarks (Fig. 3). Table Rock Lake females in the 30-60 lb. range were, on the average, 4-6 years younger than similar-sized females from Lake of the Ozarks. Additionally, similar-sized males from Table Rock Lake averaged 5-8 years younger than males from Lake of the Ozarks. Although fish of both sexes from Table Rock Lake were heavier for a given age, the slopes of age-weight regression lines indicate that the rate of growth per year for females from both lakes was similar (F=0.0074, df 1,53). Males from Table Rock Lake grew at a faster rate than males from Lake of the Ozarks (F=4.74, df 1,215). The large amounts of fatty tissue in the body cavities of Table Rock Lake males is another indication of their excellent condition.

Length-weight curves for these same paddlefish indicate that there was no significant difference between the slopes of the regression lines (Fig. 4. — Females: F=1.05, df 1,99; Males: F=0.56, df 1,449). This indicates that similar-sized males and females from Table

Rock Lake and Lake of the Ozarks are gaining weight at approximately the same rate for each unit increase in length.

The excellent growth of paddlefish in Table Rock Lake reflects the high productivity of the James River and a suitable lentic environment in the lake. According to Pflieger[398], the largest paddlefish populations are found in man-made impoundments that provide favorable feeding habitat and are fed by rivers that meet their exacting spawning requirements. We are hopeful that Table Rock Lake will meet these standards.

Harvest

Fishing for paddlefish in the James River Arm of Table Rock Lake began in 1982, when the paddlefish reached an acceptable size of 30-40 lb., and fishermen became increasingly aware of areas in the James River where they could catch them. During that first year a significant harvest occurred, and about 200-300 paddlefish were caught. During 1983, more anglers began catching paddlefish and the harvest increased substantially. Although we didn't conduct a creel census, local conservation agents estimated that as many as 2,000

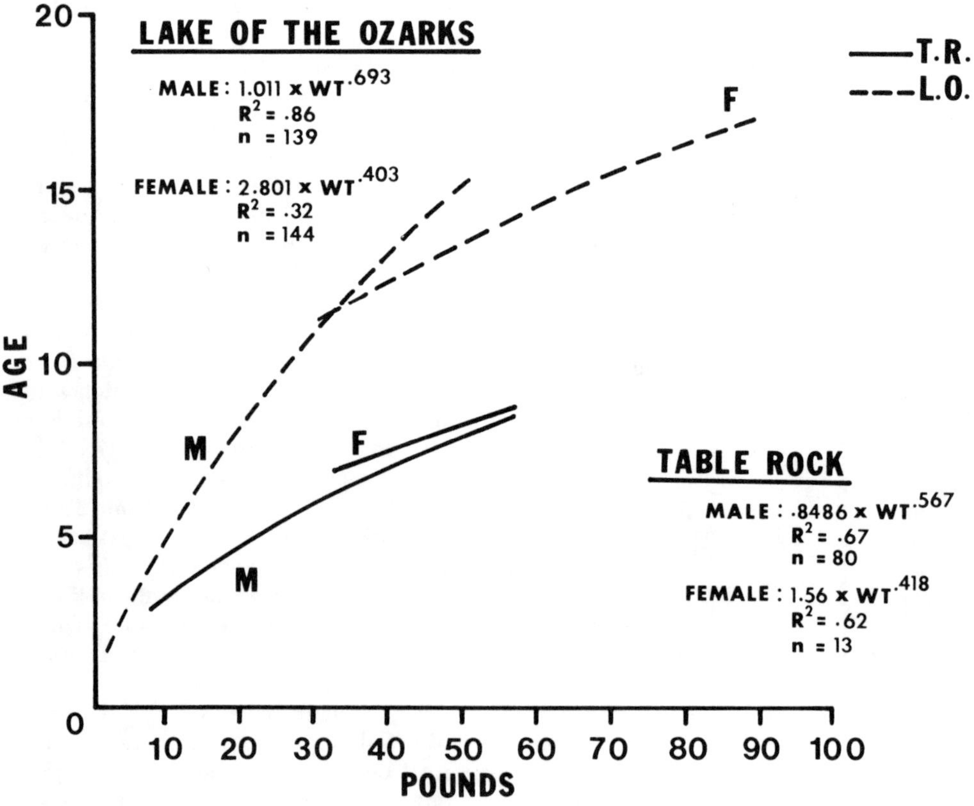

Figure 3. *Age-weight relationships for paddlefish in Table Rock Lake and Lake of the Ozarks, Missouri, 1978-1981.*

paddlefish were caught.

A creel census in the spring of 1984 indicated that an estimated 970 paddlefish weighing 39,608 lb. were harvested during the March 15–May 15 snagging season. This estimate is considered conservative because we creeled only fishermen on the lake. During the last 3 weeks of the season, several hundred paddlefish were caught upstream in the James River during a substantial rise in water levels. Fifty-six percent of the creeled fish were males averaging 33 lb. each, and the remaining 44% were females averaging 50 lb. During 1983 and 1984 these fish were harvested by snagging from boats except during substantial rises in water level. The water rose both years, and snagging from the bank became popular in several deep pools as far as 20-25 mi. upstream in the James River.

Reproduction

Most female paddlefish appear to reach sexual maturity when they are 10-12 years old and 40 lb. or larger. The paddlefish population in Table Rock Lake seems to be reaching this stage of development, because 37% of the fish in the 1983 net catch and 35% in the 1984 net catch weighed over 40 lb. (Fig. 2)

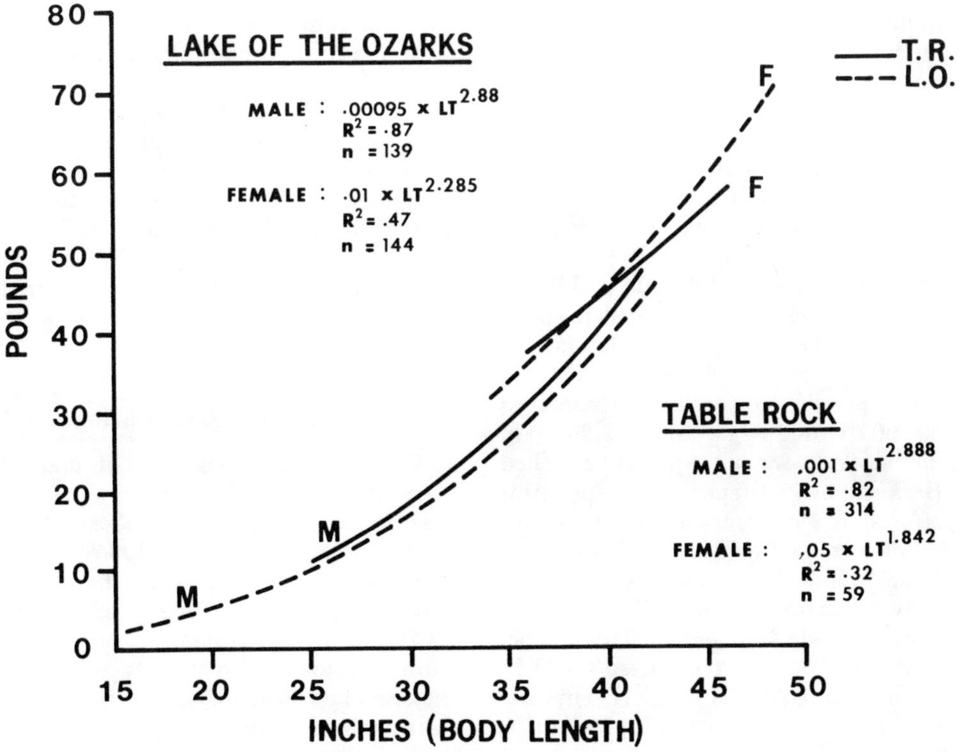

Figure 4. *Length-weight relationships for paddlefish in Table Rock Lake and Lake of the Ozarks, Missouri, 1978-1981.*

and ranged from Age VII to X. By comparison, only 10% of the population sampled by netting in 1980 was composed of paddlefish 40 lb. and larger. Some of the Age IX and X females have reached sexual maturity, because we found three fertilized eggs in samples from 1/2-m (19.6 in.) diameter tow nets after a rise on the James River during the spring of 1983.

In 1982 and 1983, 30 females that appeared to be gravid (all exceeded 45 lb.) were captured in the James River Arm of Table Rock Lake, transported to the Missouri Department of Conservation's Blind Pony Hatchery near Sweet Springs, Missouri, and injected with hormones to determine if viable eggs could be collected. Only a few viable eggs were collected each year, indicating that only a small percentage of the Table Rock Lake females were mature, even though most of these fish were larger than the minimum size of mature females from Lake of the Ozarks. Females from Lake of the Ozarks are mature when they reach Age X to XII and weigh over 40 lb. (Russell[450]).

In 1984, 11 gravid female paddlefish (47-59 lb. each) were collected from the James River Arm of Table Rock Lake and injected with hormones. Four of these females released approximately 500,000 viable eggs, and 272,000 paddlefish fry were subsequently hatched. Because of our successful spawning efforts in 1984, our entire production needs were met using these Table Rock Lake brood stock. Based on age and growth information collected from similar-sized females in Table Rock Lake in 1984, most of these 50-60 lb. females were Age XI and XII. This was similar to the age for sexual maturity of Lake of the Ozarks females, although fish in Table Rock Lake were slightly heavier for a given age than were those from Lake of the Ozarks. We are hopeful that a high proportion of the large females in Table Rock Lake are now mature and will spawn when environmental conditions in the James River permit.

Natural reproduction in the James River was not documented until 1983. Although paddlefish may not have spawned naturally for other reasons, the James River was extremely low during the springs of 1980-1982, and paddlefish could not move upstream over gravel barriers. In 1983, the James River crested at about 10 ft. over flood stage in early May. Water temperatures were ideal for spawning, and some paddlefish moved nearly 50 mi. upstream. During the 7 to 10-day rise in water levels, we collected a few fertilized eggs from gravel areas 10-25 mi. upstream from the lake.

In 1984, the James River crested at about 15 ft. over flood stage in late April, but water temperatures never exceeded 55 F. Some paddlefish moved upstream 20-25 mi. but most fish remained in the lake. Our sampling of gravel areas did not indicate that paddlefish spawned during 1984. Cool water probably prevented natural reproduction, although flows were nearly ideal.

Management Implications

The substantial harvest of paddlefish from the James River Arm of Table Rock Lake during 1983 and 1984 indicated that Table Rock Lake paddlefish were concentrating in the James River during the spring and were available to fishermen. Fishing pressure increased dramatically when the fish reached 30-40 lb. and when anglers learned how to catch them.

We believe that our initial stocking rate of about two paddlefish per acre was sufficient to establish a population capable of supporting a fishery. If these paddlefish do not reproduce naturally, supplemental hatchery stockings will be made in order to maintain a population capable of sustained annual harvests.

Hatchery production of paddlefish is expensive and not without risk (Graham[220]). Most procedures used in collecting eggs and rearing fingerlings require personnel with considerable amounts of fish culture experience, and need further work before they become common cultural practices. A dependable brood stock source must be located because gravid females are difficult to transport. In addition, a well-equipped hatchery is necessary to ensure that adequate numbers of paddlefish can be reared.

The size of paddlefish selected for stocking is an important consideration. In general, wild adult paddlefish are not economically suited for stocking because their availability is limited. Adult stockings would be practical only if natural reproduction is expected, in which case adult stockings would reduce the 7-12 year wait while the fingerlings mature. Paddlefish fry would have the best chance for survival when stocked in lakes that are new and filling, or in waters where predator fish populations are low. Although paddlefish grow relatively fast from fry to fingerling size, they are pelagic and cannot escape predation unless the water is turbid. We have found in Missouri that our best stocking results are achieved using advanced fingerlings, 10-12 in. TL.

Paddlefish have precise spawning requirements. They spawn only in flowing water over a gravel substrate when the proper combination of photoperiod, water temperature, and water flow occurs (Russell[446]). The short duration of the rise on the James River in 1983 (about 10 ft.) probably explains why more paddlefish did not spawn successfully. The substantial rise in 1984 (about 15 ft.) was probably too cool (55 F) for natural reproduction.

Paddlefish can be stocked into reservoirs lacking necessary tributary streams for natural reproduction, and harvested once they grow to an acceptable size. If the lake has a tributary stream, even one not suitable for natural reproduction, paddlefish will concentrate near its confluence during the spring and can be captured by sport or commercial methods. Although paddlefish do well in small lakes, such lakes cannot support enough fish to be profitable; therefore, larger lakes are preferred for stocking.

Although paddlefish have proven difficult to culture, there is more interest each year from commercial fish producers in rearing paddlefish either intensively or in polyculture with channel catfish (*Ictalurus punctatus*) (Semmens and Shelton[456]). Most of these private fish farmers want to culture the fish for food purposes. However, Pasch and Alexander[390] report a rapidly growing interest in paddlefish caviar. Since paddlefish can be trained to accept an artificial diet, they can be reared intensively in tanks or grown in heavily fertilized ponds. When stocked in polyculture, fertilization is generally not needed because waste food stimulates adequate zooplankton production.

Establishing paddlefish populations appears to have a bright future in sport and commercial management strategies. Reservoir and lake stocking on a large scale was first attempted in Missouri in the early 1970s, and since then several other states have begun stocking experiments. Auburn University in Alabama stocked a 350-a. lake with 440 1-lb. paddlefish in 1983. They plan to commercially harvest these fish once they have grown large enough to be used for food. The Kansas Game and Fish Commission stocked about 5,000 fingerlings (4-20 in. TL) into 9,000-a. John Redmond Reservoir from 1981-1983. Biologists from Kansas are hopeful that this population will develop and eventually spawn over gravel areas in the Neosho River upstream from the reservoir.

From 1974-1977, approximately 330,000 paddlefish fry were sent from Missouri to the USSR to be tested as a food fish and as a future source of caviar. Recently Japan and Egypt have expressed a desire to stock paddlefish as a food fish, and to establish populations in large river systems such as Egypt's Nile River. Considering the limited availability of paddlefish fry in the United States, the foreign demand for these fish will soon exceed supply.

Missouri is currently stocking fingerling paddlefish in Lake of the Ozarks to maintain that fishery, since Truman Reservoir inundated all known spawning areas in the Osage River Basin. In addition, fingerlings are being stocked into Truman Reservoir to establish a population.

Results from initial and supplemental stockings of paddlefish are very encouraging. We believe that these management techniques will enable fishery managers to develop, maintain or expand paddlefish populations in the face of ever-present environmental threats. If dedication and commitment are a part of management goals, then sport and commercial paddlefish populations can prosper in the United States.

References

DIEFFENBACH, W. H. AND F. M. RYCK, JR. 1976. Water quality survey of the Elk, James and Spring River basins of Missouri, 1964-1965. Aquatic Series No. 15. Missouri Department of Conservation, Jefferson City, Missouri, USA. 25 pages.

Opportunities in Paddlefish Aquaculture

Kenneth J. Semmens and William L. Shelton[1]

Department of Fisheries and Allied Aquaculture
Auburn University
Auburn, Alabama 36849

Abstract

Paddlefish (*Polyodon spathula*) may be produced by either intensive or extensive aquaculture. However, limited experience with the techniques of production and poorly developed markets are major factors limiting success in this industry. Markets for paddlefish roe to make caviar are excellent, but markets for paddlefish flesh are yet to be developed.

Paddlefish have potential in intensive aquaculture since they integrate well into catfish production. Fingerlings can be reared and harvested from primary catfish nursery ponds before these ponds are stocked with catfish fry. Yields from properly managed ponds should average 110 lb./a. and average survival should exceed 30% after 40 days. Paddlefish fingerlings in polyculture with channel catfish (*Ictalurus punctatus*) in secondary nursery ponds may grow to stocker size (1/3 to 1 lb.) with an expected average survival of 75%. Over 90% of these may be selectively harvested from the channel catfish fingerlings using a large mesh seine. Stocker paddlefish (400/a.) grown with fingerling catfish (3,000/a.) contributed 480 lb./a. to the harvest without significantly affecting catfish yield.

Paddlefish may also have potential for extensive aquaculture in lakes and reservoirs, although this technique remains untried. Paddlefish grow rapidly, are native to most of the United States, are easily harvested with large-mesh (6-9 in. bar-mesh), loosely hung gill nets, and are rarely captured by sport fishermen on a lure or baited hook. Extensive aquaculture of paddlefish, or reservoir ranching as we propose it, would be a system of production where paddlefish are stocked in a reservoir as fingerlings, and harvest rights to mature fish are bid out by the regulating agency to a commercial fisherman.

Raising paddlefish (*Polyodon spathula*) commercially has been possible only in the last 10-15 years with the development and refinement of propagation techniques. Paddlefish culture efforts actually began in 1915 (Alexander[10]); however, Purkett[410] was the first to successfully hatch and raise them although others (Meyer and Stevenson[329]) had tried. Purkett's work in Missouri was continued by Needham[352] and Russell[444], and the techniques they developed have been used by others in South Dakota (Unkenholz[520]; Brandt[63]), Alabama (Semmens[455]) and Tennessee (personal communication, 1983, Ronald W. Pasch, Biologist, Tennessee Valley Authority, Knoxville, Tennessee). Research at Auburn University coupled paddlefish spawning methods with those for white sturgeon (*Acipenser transmontanus*) described by Doroshov et al. (1983). As a result, it is possible to consistently produce paddlefish on a commercial basis.

[1]Present Address: Zoology Department, University of Oklahoma, Norman, Oklahoma 73019

Today fairly reliable and predictable techniques are available to stimulate ovulation, hatch eggs and rear paddlefish (Graham[220]). Although refinements in some techniques are needed, the basic process has been worked out.

Factors now limiting paddlefish aquacultural opportunities are markets and marketing. The potential appears good, but much remains to be done before these markets permit a significant industry.

In this paper we examine the markets for paddlefish, review what is known about raising them on a commercial basis and discuss some opportunities for the future.

Markets

Paddlefish were first captured and sold as food in the United States in the late 1800s. Before then, a prevailing popular prejudice against the species, and a readily available and abundant supply of sturgeons, contributed to low use of paddlefish. Sturgeons were more quickly accepted because they had long been considered valuable food fish in Europe, and were more familiar to early immigrants of European descent.

Today, the market for paddlefish products remains relatively limited and unorganized except for caviar. A few paddlefish are sold for meat, for stocking, as specimens for biological supply companies, or as pets or curios in aquariums and museums. Unfortunately, it has been extraordinarily difficult to ship fingerling paddlefish to overseas aquariums without massive mortalities. Other domestic and international markets remain largely unexplored due to lack of a consistent supply of paddlefish.

Social and legal limitations may limit marketing of paddlefish. Many states classify paddlefish as a sport fish, which places restrictions on the sale of paddlefish or paddlefish parts. As a result, tens of thousands of dollars worth of roe are discarded by sport fishermen. Even stricter limitations exist in Wisconsin, Texas, Ohio and Minnesota, which consider paddlefish a threatened species and prohibit any harvest.

Alabama, Mississippi, Louisiana, Tennessee and Arkansas classify paddlefish as a commercial species. However, except for caviar, paddlefish products generally have low value in their markets. Still other states (i.e. Georgia and California) consider paddlefish an exotic species and will not permit commercial culture in any form.

Throughout its natural range, paddlefish is seldom sold under its own name. The flesh is usually sold as boneless catfish or some sort of sturgeon, and caviar is generally marketed as "American sturgeon." Paddlefish is a fine smoked fish and has recently appeared in commercial smokehouses.

Caviar

Harvest of paddlefish eggs to make caviar first occurred in the Mississippi Valley about 1896 (Townsend[511]) and became an important industry during the early 1900s. Caviar production soon switched back to Europe, however, and it was not until the recent Iranian hostage crisis that interest in paddlefish eggs for caviar was revived. Today, the market for eggs is the best developed paddlefish market, and the one most accessible to commercial fishermen.

Paddlefish caviar is generally processed by the person who captures the fish. Fishing usually begins in late November and continues into the spawning season. Ovaries are taken from gravid females and gently pushed through a 1/4-in. wire mesh screen to separate the eggs from surrounding membrane and fatty tissue. Screened eggs are then rinsed in a light salt solution.

The cleaned eggs are again salted, then drained. The amount of salting and draining determines the saltiness and dryness of the final product. Color varies from dark black to light gray, but frequently has a greenish or brownish tinge. Egg size is comparable with sevruga caviar.

Fishermen usually sell their raw caviar to buyers who in turn sell it to established caviar companies, mostly in the New York area. Price varies with quality, season, and perceived gullibility of the seller, but buyers frequently pay up to $30/lb. for the minimally processed eggs. At this time paddlefish caviar is available only from natural populations. Demand will likely increase as natural stocks of paddlefish are depleted or closed to commercial harvest.

Flesh

Currently, marketing paddlefish flesh is difficult because the supply is undependable and the quality is highly variable. The yearly commercial harvest of paddlefish (approximately 286 t.) is less than the estimated annual sport harvest of 307 t. (Carlson and Bonislawsky[95]). Paddlefish caught in summer yield an inferior product, because they quickly die when captured in gill nets during warm weather. Furthermore, intestinal parasites begin to migrate from the anus at death and present a most unappealing sight. Harvest in winter is frequently associated with demand for caviar, and at that time flesh is of secondary importance.

Unlike catfish, paddlefish are usually bled and the lateral bands of dark meat removed to improve product appearance and flavor. There is considerable waste, and only about one-third of the fish remains as white, boneless, skinless fillets. In the market, paddlefish are usually sold under another name, such as boneless catfish or sturgeon. As a result, markets for paddlefish flesh are poorly developed, with prices comparable to dressed buffalo (25-35 cents/lb.). Historically, dressed paddlefish have been valued from one to five times that of dressed buffalo (Carroll[97]; Coker[106]).

Paddlefish flesh is firm, lean and boneless with a beef or pork-like texture. It is most tasty baked, broiled or smoked but can also be boiled or fried. Smoked paddlefish substitutes directly for sturgeon, producing an excellent product. As a result, larger fish are preferred by smokehouses.

Aquaculture offers opportunities for paddlefish production similar to those that have been developed for catfish: uniform product of good quality at a reasonable, consistent price in predictable and consistent supply. Paddlefish should complement rather than substitute for channel catfish (*Ictalurus punctatus*) products, permitting needed product diversity. Prices for cultured paddlefish will have to be greater than for both wild paddlefish and catfish products, but less than for cultured sturgeon. Products from sturgeon reared in raceways will be relatively expensive because they must be fed commercially prepared diets (personal communication, 1983, Ken Beer, The Fishery, Sacramento, California). Other sturgeon products such as oil and isinglass should be investigated as potential paddlefish products.

Methods of Production

Although possible, paddlefish production requires the talents of a skilled fish culturist (Graham[220]). To date only a handful of aquaculturists have consistently hatched and reared significant numbers. We believe paddlefish may be cultured on a broader scale both intensively and extensively, although the latter is discussed more as a proposal than a proven technique.

Intensive Aquaculture

More information is available on paddlefish fingerling production under intensive conditions than for other production systems (Graham[220]). All steps necessary for collecting brood stock, injecting females to induce spawning, hatching eggs, preparing ponds and raising fingerlings have been used and improved by the state fisheries workers in Missouri. Although their production has been consistent, they point out that further refinements are necessary before large-scale production of fingerlings is considered routine. They do not, however, outline procedures for rearing paddlefish to adult size to satisfy other potential markets.

Outside of state and federal agencies, we know of only two individuals who have gained the experience and knowledge necessary to produce paddlefish. One produces young paddlefish in ponds by themselves (monoculture); we have raised them with other species (polyculture).

Polyculture is a logical means of diversification in freshwater aquaculture and is probably the best way to introduce paddlefish into aquaculture. Paddlefish are more desirable for polyculture systems than many other species, because they are able to utilize natural foods and not compete with the primary species. Also, paddlefish are native to most of the United States and are not, therefore, restricted by importation regulations as are many other species well-suited for polyculture. Paddlefish are capable of rapid growth to a large size (Table 1) and do not compete with channel catfish in rearing ponds (Kirkendall[283]).

One consideration when using paddlefish in a polyculture system is that

Table 1. *Potential growth of paddlefish at low density in a moderately fertile pond environment.*

Mean weight (kg)		Growth period (days)	Growth rate (g/day)	Average daily weight increase (% initial weight/day)	Source
Stocking	Harvest				
0.00005	0.009	33		540	1
0.009	1.2	123	7.7	107	2
0.32	2.75	120	20.5	6.32	3
1.0	3.8	225	12.4	1.24	4
4.1	8.4	114	37.3	0.92	5
5.1	12.0	144	47.9	0.94	5
8.6	27.2	596	31.2	0.36	6

1. Semmens[455]
2. Busch and Shell[89]
3. Semmens (unpublished data)
4. Busch[88]
5. Swingle[493]
6. Shelton and Meriwether[460]

they are filter feeders, and can utilize only zooplankton and insects larger than 0.01 in. diameter (Rosen and Hales[427]). They do not efficiently utilize phytoplankton, and therefore do not have the stabilizing effect upon oxygen concentrations that has been attributed to silver carp (*Hypophthalmichthys molitrix*) (Piotrowska 1978).

Our experience has shown that paddlefish grow-out should be coupled with catfish grow-out in intensive aquaculture. Feeding catfish results in direct and indirect pond fertilization which enhances the pond's carrying capacity for fishes relying on natural foods. This abundance of natural foods may be used to produce paddlefish without increasing feed costs. Of course paddlefish are stocked at a relatively low density compared with channel catfish, and the paddlefish have to be thinned as the pond's capacity is reached. Table 2 outlines appropriate stocking densities, assuming 90% annual survival in the grow-out phase and a carrying capacity of 450 lb./a.

Paddlefish rely on natural foods throughout this culture cycle, so overstocking should be avoided. Even intensive culture of paddlefish fingerlings requires considerable space (personal communication, 1983, Jerry Hamilton, Fish Hatchery Manager, Blind Pony Hatchery, Sweet Springs, Missouri). Because they must keep moving to obtain oxygen, paddlefish cannot be crowded like catfish. As a result, in polyculture they *must be* separated from catfish during harvest.

Paddlefish are easily captured by seining. When a seine is being pulled in, catfish push against the net; but paddlefish shun the net and swim to open water above the catfish. As the net is pulled tighter, paddlefish swim to the surface and tire easily. When exhausted they roll over and float. They must be moved quickly to a holding pen adjacent to the haul seine, or they will soon die without adequate space to swim. Paddlefish are easily captured by the rostrum or tail and hand-sorted from the catfish.

Proper management of paddlefish in intensive aquaculture necessitates an understanding of the relationship between carrying capacity, potential growth, and stocking density. Carrying capacity for paddlefish in a pond may

Table 2. *Expected production of paddlefish in polyculture with channel catfish, assuming 450 lb./a. carrying capacity, in a 1-a. pond.*

Age group	Stocking rate (No./a.)	Harvest rate (No./a.)	Average stocking size (kg)	Average harvest size (kg)	Growth period (days)	Average daily weight increase (%/day)
0	30,000	10,000	0.00005	0.005	40	248[1]
0	600	450	0.005	0.45	180	49[2]
1	100	90	0.45	2.3	365	1.13[3]
2	42	38	2.3	5.4	365	0.37[4]
3	20	18	5.4	11.4	365	0.30[4]
4	16	15	11.4	15.9	365	0.11[4]

[1]Fry in monoculture primary nursery ponds
[2]Secondary nursery ponds — fingerlings in polyculture with catfish fry to fingerlings
[3]Growout ponds in polyculture with channel catfish "stockers"
[4]Growout ponds in polyculture with channel catfish foodfish

be defined as the weight of paddlefish that can be produced in a given area. An infertile reservoir may be capable of supporting only 10 lb./a. of paddlefish. In contrast, in a catfish pond where the catfish are fed daily and water fertility is enhanced by adding organic matter, approximately 500 lb./a. of paddlefish can be produced. Table 1 outlines the rapid growth potential for paddlefish at relatively low densities in a moderately fertile pond. However, under actual production conditions the growth might not be as rapid (Table 2) because of competition with other fishes and poor water quality, especially low dissolved oxygen. Catfish can survive low levels of dissolved oxygen that will stunt or kill paddlefish.

Paddlefish fingerling (age group 0) production consists of a primary and secondary nursery phase. The primary phase, rearing the fry, is completed before catfish spawn, and does not conflict with use of nursery ponds for catfish fry. Although yields may exceed 200 lb./a., and survival may average 50% (Semmens[455]), a more realistic yield under protection conditions would be 110 lb./a. with an average survival of 30% after 40 days.

In the secondary phase, paddlefish fingerlings may be grown in polyculture with catfish fry or fingerlings. Survival during the secondary phase is expected to average 75%. Size at harvest depends on stocking density, pond fertility and water quality. Harvesting with a large-mesh seine will allow catfish fingerlings to escape while selectively retaining stocker-sized paddlefish. A single seine pass will usually capture over 90% of the paddlefish. Fingerlings that remain mixed with catfish are easily hand sorted.

Growing paddlefish for caviar in an intensive system is risky and more costly than producing fingerlings. Therefore, intensive culture of paddlefish would be best oriented toward production of flesh instead of roe. Paddlefish could be harvested after the second year, or when the fish reach 5 lb. In this way, several crops of paddlefish could be produced for flesh in the same time required to produce a single crop of caviar.

Extensive Aquaculture

To our knowledge extensive aquaculture of paddlefish is untried, although it could result in substantial numbers for either flesh or caviar markets. Extensive aquaculture, or — as we propose — reservoir ranching, is a system of production where paddlefish are stocked in a body of water and limited harvest rights are sold to commercial fishermen.

Paddlefish seem particularly well-suited for this type of production management. They grow rapidly, reach large sizes, are long-lived and are fairly easy to capture.

Natural production — Because paddlefish are zooplanktivores, one would expect standing stocks to be comparable in size to those of other fish in this feeding niche. For instance, the standing stock of bighead carp (*Aristichthys nobilis*), another zooplanktivore, was about 22 lb./a. in unfertilized reservoirs (Yang 1970) and 52 lb./a. in unfertilized ponds in Taiwan (Lin 1968; Shelton and Smitherman[461]). In the United States, Combs[112] estimated there were 25,000 harvestable-size paddlefish in Grand Lake of the Cherokees, Oklahoma (51,900 a.) averaging about 42 lb., or almost 20 lb./a. Alexander and Peterson[9] estimated there were approximately 8,700 harvestable-size paddlefish in Norris Reservoir, Tennessee (34,000 a.) or about 10 lb./a. We believe that standing stocks of harvestable-size paddlefish could be greater than 10-20 lb./a. with increased fertility, recruitment or stocking.

Reservoir ranching — In its simplest form, reservoir ranching would require a contract between an aquaculturist and the agency regulating the reservoir. The aquaculturist would be allowed to stock the reservoir with fingerling paddlefish, and in exchange for compensating the agency, the aquaculturist would have exclusive rights to harvest mature paddlefish. Precursors to this system are in operation today. The Missouri Department of Conservation established a paddlefish population in Table Rock Lake, Missouri, with hatchery-reared fingerlings (Graham[220]). This population is harvested only by sport fishermen. In Tennessee's Cherokee Reservoir, commercial fishermen harvest naturally occurring paddlefish on a contract basis with the state agency (personal communication, 1983, David Bishop, Fisheries Biologist, Tennessee Wildlife Resources Agency, Morristown, Tennessee). Reservoir ranching would merely combine the main elements of these two production systems.

Large paddlefish can be harvested in reservoirs with large-mesh, loosely hung gill nets. Because large paddlefish swim constantly, often congregate, and are relatively easy to locate with electronic fish finders, they are likely to encounter a net.

Even in large reservoirs, gill nets can be quite effective (Carroll[97]). Commercial fishermen harvested 87% of their total catch of paddlefish within the first 3 weeks of the season in Cherokee Reservoir, Tennessee, in 1982. However, the most impressive example of gill net effectiveness was in Bard's Lake, Tennessee. In two nights, two fishermen harvested over 12,000 lb. of paddlefish from a 320-a. subimpoundment, and captured only one game fish (Jenkins[262]). Incidental catches of game fishes can be minimized by requiring the use of 8 to 9-in. bar-mesh gill nets. Harvest rates from previously unfished reservoirs are summarized in Table 3.

Problems and challenges — There are several potential complications in reservoir ranching. Illegal fishing by non-contract fishermen could reduce profit. Emigration of stocked fingerlings could also affect yields and profit. In reservoirs such as Norris Reservoir, Tennessee, the potential for incidental

Table 3. First season harvest of paddlefish from unfished waters.

Impoundment	State	Surface area (a.)	Harvest (lb./a.)	Harvest duration (days)	Reference
Norris	Tn	34,000	1.3	30	1
Wheeler	Al	67,000	1.7	240	2
Barkley	Tn/Ky	58,000	2.3	150	3
Wilson	Al	16,000	37	240	2
Lake Energy	Ky	350	31	2	4
Bard's Lake	Tn	320	37	2	4

1. Carroll[97]
2. Bryan[76]
3. Carlson and Bonislawsky[95]
4. Jenkins[262]

capture of trophy striped bass may cause conflict with sport fishermen. Paddlefish harvest by sport fishermen could be a problem if suitable fishing areas exist and environmental conditions are especially favorable. Introduction of paddlefish into new systems may conflict with present fisheries in some reservoirs. However, if this became an issue, stocking rates could be adjusted to minimize competition.

Reservoir ranching represents a real challenge to fishery managers, especially if conflicts develop with sport fishermen. On the other hand, through close cooperation between sport and commercial fishermen and the agency regulating the reservoir, reservoir ranching could lead to a new aquacultural enterprise.

Conclusions

Interest in commercial production of paddlefish has increased recently because of breakthroughs in artificial propagation and the development of a market for paddlefish eggs to make caviar. Although the interest is high, few commercial growers are currently producing paddlefish. We expect that more growers will become involved as wild stocks are depleted, as markets become further developed, and as the technology of production becomes more widely known and available.

At this point, the key to success or failure, to problem or opportunity, in paddlefish aquaculture lies not in production methods, which are now being learned, practiced and even improved, but in markets and marketing. Without a good, reliable marketing structure, paddlefish aquaculture will remain a relatively low-volume, high-risk business.

References

DOROSHOV, S. I., W. H. CLARK, JR., P. B. LUTES, R. L. SWALLOW, K. E. BEER, A. B. MCGUIRE AND M. D. COCHRAN. 1983. Artificial propagation of the white sturgeon (*Acipenser transmontanus*). Aquaculture 32(1-2):93-104.

LIN, S. Y. 1968. Pond fish culture and the economy of inorganic fertilizer application. Chinese-American Joint Commission on Rural Construction, Fisheries Series No. 6 Taipei, Taiwan.

PIOTROWSKA, W. 1978. The influence of silver carp (*Hypophthalmichthys molitrix*) on eutrophication of the environment of carp ponds. Part I: Physico-chemical conditions. Roczniki Nauk Rolniczych 1978, Seria H.T. 99 Z.2. Pages 7-32.

YANG, Y.A. 1970. Evaluation of balance between fishes and available fish foods in multispecies fish culture ponds in Taiwan. Transactions of the American Fisheries Society 99(4):708-718.

Symposium Wrap-up

The Editors[1]

The theme of this symposium was in the form of a question: Is the paddlefish a threatened resource? We can start our search for the answer with a quick review of the information presented by our panel of experts.

It's clear that our understanding of the paddlefish has vastly improved since Mr. Purkett's discovery of the first paddlefish eggs in 1960. We now have considerably more information on all aspects of paddlefish biology — including food habits, growth, movement and reproduction. We also know far more about paddlefish distribution, abundance and population fluctuations, since many states in the fish's range have made efforts to monitor their populations. In addition, we have developed techniques to artificially propagate the species.

Along with our increased knowledge of the biology and distribution of paddlefish, we have become more aware of how this species has been affected by human impact. We now know that dams and other water development projects have caused incremental, detrimental changes in habitat, which have lead to the gradual decline of paddlefish in many river systems. We have also seen how drastically the commercial exploitation of populations for roe can reduce a fishery. Regulation and habitat protection have been successful in some areas, and are needed in many others.

With this information in mind, let's look at the bright side of our original question. What are the facts that suggest the paddlefish is *not* a threatened resource?

Certainly the most positive evidence in this regard is the fact that, in many areas, paddlefish populations seem to be stable or increasing. Although some populations have declined and others have been lost, especially on the periphery of the fish's range, in most states the species is not considered in danger of extirpation. Even populations once decimated by commercial fishing have recovered in some areas.

Two more reasons for optimism are the recent development of methods for artificial propagation, and the successful stocking of hatchery-reared paddlefish in previously unoccupied habitat. These techniques show great promise for restoring depleted populations and could make commercial production a reality.

Another important breakthrough has been the preservation and protection of spawning habitat in the Upper Missouri and Yellowstone rivers. In these cases habitat was protected specifically for the paddlefish, indicating that scientific and public interest in this species has become great enough to have political influence.

All of this is good news — but we shouldn't become complacent. Many challenges in managing our paddlefish resources lie ahead; if they are not met,

[1]The wrap-up at the symposium was presented by Dr. Richard O. Anderson, Director, National Fishery Center, San Marcos, Texas, 78666. This text was prepared by the editors from his notes.

the question posed as the theme might still one day be answered with a YES. The paddlefish could easily become a threatened resource — threatened by inadequate management, habitat loss and commercial overexploitation.

Inadequate or poor management is a real threat. Most harvest regulations, for example, must be considered experimental at this point. Although 14 of 22 states in the fish's range regulate sport or commercial harvest, only four reported that the effects of their regulations were known. Another problem is that the creel limit, widely used in paddlefish management, has not been evaluated for its effectiveness. Regulations as major management tools are particularly questionable where the status of the paddlefish population is not well known. It is unfortunate that many states with little knowledge of their populations have classified the paddlefish as a commercial species, a status which usually allows a greater harvest than is desirable to effectively protect and manage the population. The recent increase in roe prices could compound such a situation, causing increased fishing pressure on an already overtaxed resource, ultimately resulting in a long-lasting impact on paddlefish populations similar to that which occurred in the late 1800s.

Habitat loss is still a critical threat, and one not likely to be solved without coordinated, long-term efforts. The negative effects of large reservoirs on reproduction and movement of paddlefish have been well-documented. A number of water development projects threaten the paddlefish or its habitat and the outcome is still uncertain. Pollution and stream alteration also continue to take their toll.

Commercialization of the paddlefish may or may not be a threat. On one hand, it can result in renewed interest in better management; on the other, it may increase pressure on populations and encourage illegal harvest. If roe prices remain high, fisheries managers will have to monitor populations closely to prevent overexploitation and resultant major resource losses. Commercial fishermen could help in such efforts by reporting illegal activities.

In summary, we have learned from this symposium that paddlefish populations seem to be holding their own in many areas, that public and professional interest is increasing, and that new propagation and stocking techniques hold promise for restoring populations and offer new aquacultural opportunities. However, loss of critical spawning habitat has been widespread, and is likely to continue. Commercial demand for roe is high. Harvest regulations and many other management approaches are still unproven. More importantly, many states do not have enough knowledge of their paddlefish resource to identify problems or to develop the best possible management strategies.

The answer to our original question, "Is the paddlefish a threatened resource?" appears to be neither yes nor no but NOT YET. What this means is that we are in the enviable position of having defined a major biological-ecological-political problem before it becomes a crisis. In the history of fisheries management, such opportunities are all too infrequent.

Now the question is: Will we take advantage of this opportunity? We could continue managing and studying paddlefish populations as we are now, since they don't seem to be in any immediate danger. However, such a course of action hardly seems the best use of our new awareness from this symposium. Instead, we offer the following suggestions to ensure the future of paddlefish populations in the United States.

First, we recommend that the responsible agencies consider a regional approach for managing paddlefish populations. Paddlefish travel great distances, and what is done or not done in a watershed or river system in one state may very well affect conditions in several states. Communication and understanding are especially important in coordinating harvest regulations and protecting critical habitat. In particular, more spawning sites should be identified. Destruction of unidentified spawning sites could become the most serious threat to our paddlefish resource. Since the species is highly mobile, several populations could be dependent upon one spawning area, and would suffer irrevocable losses if that particular area was destroyed.

Our second recommendation is that the individual states, especially those where the paddlefish is classified as commercial, look more closely at the status of their populations. Data on numbers, sizes and locations of populations remain sketchy in many areas. Multi-state or multi-agency research efforts could be beneficial in obtaining much-needed information for future management strategies.

Thirdly, we challenge appropriate agencies to develop reasonable management objectives for their paddlefish resource. Without clear ideas of what you need to do and where you want to be, management programs are at best haphazard and ineffective.

Finally, a note of caution. Although techniques for producing and stocking paddlefish were presented in this symposium, we do not consider stocking to be an answer to habitat deterioration and management problems. These techniques were developed for special circumstances where stocking was the only way to maintain a population. Artificial propagation and stocking should not be used as a cure-all or substitute for wise or practical management.

Trying to solve problems by treating symptoms is expensive and ineffective. In addition, stocking could affect the integrity of paddlefish gene pools and is ill-advised until we know a lot more about the genetics of this species. A better strategy is to prevent habitat deterioration and regulate harvest so that maintenance stocking is unnecessary. As far as propagation of paddlefish for commercial sale is concerned, we consider it to be in its earliest stages, and anyone interested in such a venture should be aware that markets (except those for roe) are not yet developed.

This symposium was organized for two purposes. The first was to assemble, synthesize, present and discuss, for the first time, the vast amount of information that is available on paddlefish. With the help of the symposium presentors and the symposium audience, that purpose was accomplished. The second purpose, equally important, was to encourage more effective management and protection of this intriguing and valuable fish. As symposium organizers and editors, we have less control over this objective. It will require many more years and much more effort, but we think it can be met. We solicit your help and support.

JGD
LKG
TRR
BKB

Bibliography

An Indexed Bibliography of the Paddlefish (*Polyodon spathula*)

L. Kim Graham

Fish and Wildlife Research Center
Missouri Department of Conservation
Columbia, Missouri 65201

Abstract

Five hundred and fifty-five references to technical and popular literature on paddlefish (*Polyodon spathula*) are listed alphabetically, numbered, and cross-indexed into one or more of 16 subject headings. Pertinent subject heading code numbers follow each reference. In addition, the 16 subject headings are listed, each followed by the numbers of all appropriate references. Information regarding the Chinese paddlefish (*Psephurus gladius*) is included in this bibliography because of systematic similarities between it and the North American paddlefish (*Polyodon spathula*).

This listing of technical and popular literature on paddlefish (*Polyodon spathula*) was compiled to serve as a ready reference for fishery managers, researchers, and individuals interested in this species. An earlier version (Graham and Bonislawsky[219]) served as the basis for this listing, which includes an additional 220 references.

Many of the references were obtained by contacting fisheries biologists and managers from state and federal wildlife agencies involved in paddlefish studies. Others were located by searching the Fish and Wildlife Reference Service, various biological abstracting services such as Sport Fisheries Abstracts, Biological Abstracts, and Chemical Abstracts, and scientific journals and textbooks from personal libraries. Many of the older references were located in Linda Hall Library, a research library at the University of Missouri, Kansas City, Missouri.

Acknowledgments

This bibliography is a result of close association and cooperation between state and federal fisheries biologists, and other researchers who have studied or are now studying and managing paddlefish populations. I wish to thank the following persons for their contributions to this bibliography: Mike Alexander, Emil Berard, Pat Bonislawsky, Doug Carlson, Roger Copper, Al Elser, Tom Gengerke, Todd Georgi, Don Hales, Jerry Hamilton, Don Helms, Larry Kallemeyn, Dan McKinley, Ron Pasch, Lee Rupp, Tom Russell, and Dennis Unkenholz.

References

1. Ackerman, G. L. 1975. Iowa's first snagging season for the paddlefish. Iowa Conservationist 34(2):6-9. (7,8)
2. Acton, R. T., P. F. Weinheimer, H. K. Dupree, T. R. Russell, M. Wolcott, E. E. Evans, R. E. Schrohenloher, and J. C. Bennett. 1971. Isolation and characterization of the immune macroglobulin from the paddlefish, *Polyodon spathula*. Journal of Biological Chemistry 246(22):6760-6769. (5)
3. Acton, R. T., P. F. Weinheimer, S. J. Hall, W. Neidermeier, E. Shelton, and J. C. Bennett. 1971. Tetrameric immune macroglobulins in three orders of bony fishes. Proceedings of the National Academy of Sciences 68(1):107-111. (5)
4. Adams, L. A. 1919. A memoir on the phylogeny of the jaw muscles in recent and fossil vertebrates. Annals of the New York Academy of Sciences 28:51-166. (3,4)
5. Adams, L. A. 1942. Age determination and rate of growth in *Polyodon spathula*, by means of the growth rings of the otoliths and dentary bone. American Midland Naturalist 28(3):617-630. (6,13)
6. Akhundov, A.-D. G., and B. M. Mednikov. 1976. Systematic position of the cartilaginous ganoid *Polyodon spathula* (Walbaum), (Pisces, Chondrostei). Doklady Akademii Nauk SSSR, Seriya Biologiya 228(1-6):271-272. (2,6)
7. Alexander, C. M. 1985. An assessment of the harvest potential of paddlefish stocks in Watts Bar Reservoir, Tennessee. Master of Science Thesis. University of Tennessee, Knoxville, Tennessee, USA. 45 pages. (6,9,10,13)
8. Alexander, C. M., and T. A. McDonough. 1983. Effect of water conditions during spawning on paddlefish year-class strength in Old Hickory Reservoir, Tennessee. Tennessee Valley Authority, Office of Natural Resources, Division of Air and Water Resources, Knoxville, Tennessee, USA. Progress Report. 26 pages. (6,10,11)
9. Alexander, C. M., and D. C. Peterson. 1982. Feasibility of a commercial paddlefish harvest from Norris Reservoir. Proceedings of the Annual Conference of the Southeastern Association of Fish and Wildlife Agencies 36:202-212. (9)
10. Alexander, M. L. 1914. The paddlefish (*Polyodon spathula*). Transactions of the American Fisheries Society 44(1):73-78. (6)
11. Alexander, M. L. 1915. More about the paddlefish (*Polyodon spathula*). Transactions of the American Fisheries Society 45(1):34-39. (6)
12. Allen, W. F. 1907. Distribution of the subcutaneous vessels in the head region of the ganoids, *Polyodon* and *Lepisosteus*. Proceedings of the Washington Academy of Science 9:79-158. (4)
13. Allen, W. F. 1911. Notes on the breeding season and young of *Polyodon spathula*. Journal of the Washington Academy of Science 1(10):280-282. (6,11)
14. Allis, E. P. 1903. On certain features of the lateral canals and cranial bones of *Polyodon folium*. Zoologische Jahrbuecher 17(4):659-678. (4)

15. ALLIS, E. P. 1911. The pseudobranchial and carotid arteries in *Polyodon spathula*. Anatomischer Anzeiger 39:257-262, 282-293. (4)

16. ALLIS, E. P. 1920. The branches of branchial nerves of fishes with special reference to *Polyodon spathula*. Journal of Comparative Neurology 32(2):137-153. (4)

17. ALLIS, E. P. 1935. On a general arrangement of the cranial roofing bones in fishes. Journal of Anatomy 69:233-291. (4)

18. AMBLER, M. E. 1984. Paddlefish exploitation. Proceedings of the Oklahoma Chapter of the American Fisheries Society annual meeting. 1(1):11-12. (6,8,9,10)

19. AMBLER, M. E. 1985. Paddlefish investigations: fish research and surveys for Oklahoma lakes and reservoirs. Oklahoma Department of Wildlife Conservation. Dingell-Johnson Project F-37-R-7, Job Number 6. Performance Report. 10 pages. (8,9,10,13)

20. ANDERSON, M., AND G. L. ACKERMAN. 1977. Fire in the hole. Iowa Conservationist 36(2):8-9. (7,8)

21. ARTEDI, P. 1788-1793. Petri Artedi renovati partes . . . (ichthyologiae). Edited by J. J. Walbaum. A. F. Rose, Grypeswaldiae. 5 Volumes. Volume 2:522. (2)

22. AUCKLEY, J. 1974. Paddle trackers. Missouri Conservationist 35(3):4-7. (7,10)

23. AXON, J. R., W. N. MCLEMORE, D. E. BELL, B. D. LAFLIN, J. P. HENLEY, L. E. KORNMAN, A. J. JONES, AND K. W. PRATHER. 1980. Statewide fisheries management project. Part I: Reservoir research and management. Kentucky Department of Fisheries and Wildlife Resources. Dingell-Johnson Project F-50-2, Part 1, Study Number 1-2. Progress Report. 150 pages. (1,6,10)

24. BACKLUND, C. A. 1970. Missouri River elder. South Dakota Conservation Digest 37(2):2-5. (7,8)

25. BAILEY, R. M., AND M. O. ALLUM. 1962. Fishes of South Dakota. University of Michigan Museum of Zoology Miscellaneous Publication Number 119, Ann Arbor, Michigan, USA. 131 pages. (1,6)

26. BAKER, C. L. 1937. The commercial, game, and rough fishes of Reelfoot Lake. Journal of Tennessee Academy of Science 12(1):9-59. (1,6)

27. BAKER, C. L., AND M. V. PARKER. 1938. The fishes of Reelfoot Lake. Journal of Tennessee Academy of Science 13(2):160-163. (1)

28. BALLARD, W. W., AND R. G. NEEDHAM. 1964. Normal embryonic stages of *Polyodon spathula* (Walbaum). Journal of Morphology 114(3):465-477. (11,14)

29. BARBOUR, R. W. 1951. Occurrence of paddlefish in eastern Kentucky. Copeia 1951(1)1:102. (1,11)

30. BARBOUR, T. 1911. The smallest *Polyodon*. Biological Bulletin (Woods Hole). 21(4):207-215. (6,11)

31. BARDACH, J. E., J. H. RYTHER, AND W. O. MCLARNEY. 1972. Early attempts at fish farming in the south central United States using buffalofishes and paddlefish. Pages 147-158 *in* Aquaculture; the farming and husbandry of

freshwater and marine organisms. Wiley-Interscience, New York, New York, USA. 868 pages. (14)

32. Barnickol, P. G., and W. C. Starrett. 1951. Commercial and sport fishes of the Mississippi River between Caruthersville, Missouri and Dubuque, Iowa. Illinois Natural History Survey Bulletin 25(5):267-350. (9)

33. Barsness, J. 1981. First Person. Sports Illustrated. June 1, 1981, 54(23): 94, 96, 98. (6,7)

34. Beach, H. 1902. The paddlefish, *Polyodon spathula*. Bulletin of the Wisconsin Natural History Society 2:85-86. (6)

35. Beard, D. C. 1878. The paddlefish of the Mississippi. Scientific American 39(25):391. (6)

36. Beaver, P. C., and P. H. Simer. 1940. A restudy of the three existing species of the cestode genus *Marsipometra* (Cooper) from the spoonbill, *Polyodon spathula* (Walbaum). Transactions of the American Microscopical Society 59(2):167-182. (15)

37. Beck, D. 1978. The ponderous paddlefish. Iowa Conservationist 37(3):2-3. (7)

38. Becker, G. C. 1976. Inland fishes of the Lake Michigan Drainage Basin. Argonne National Laboratory Reviews, Volume 17 (ANL/ES-40). Chicago, Illinois, USA. 237 pages. (1)

39. Becker, G. C. 1983. Fishes of Wisconsin. The University of Wisconsin Press, Madison, Wisconsin, USA. 1,052 pages. (1,6)

40. Beedle, D., and T. Georgi. 1977. The kidney of the chondrostean fish, *Polyodon spathula* (Walbaum). Proceedings of the Nebraska Academy of Sciences and Affiliated Societies 87:7. (4)

41. Berg, R. K. 1980. Spoonbill. Montana Outdoors 11(3):11-13, 21-22. (6,7)

42. Berg, R. K. 1981. Fish populations of the wild and scenic Missouri River, Montana. Montana Department of Fish, Wildlife and Parks. Federal Aid to Fish and Wildlife, Restoration Project FW-3-R, Job Number 1-A. 242 pages. (8,10,11,13)

43. Berg, R. K. 1985. Managing fish on the "Mighty Mo". Montana Outdoors 16(4): 6-10. (6,7)

44. Bernet, C. K. 1976. A study of fishes impinged at the Allen Steam Plant from August 1974 through February 1976. Master of Science Thesis. University of Mississippi, Oxford, Mississippi, USA. 94 pages. (1,6)

45. Bernet, C. K., Y. J. McGaha, and C. M. Cooper. 1977. Observations on the growth rate of young paddlefish. Journal of the Mississippi Academy of Sciences 22:109. (6,13)

46. Bianchi, D. R. 1969. Inventory of the waters of the project area. Montana Department of Fish and Game. Dingell-Johnson Project F-30-R-4, Job Number 1. Progress Report. 4 pages. (8,10)

47. Bianchi, D. R. 1970. Inventory of the waters of the project area. Montana Department of Fish and Game. Dingell-Johnson Project F-30-R-5, Job Number 1. Progress Report. 5 pages. (8,10)

48. Bianchi, D. R. 1971. Fish management surveys. Montana Department of

Fish and Game. Dingell-Johnson Project F-30-R-6, Job Number 1-B. Progress Report. 5 pages. (8,10)

49. Black, D. 1917. The motor nuclei of the cerebral nerves in phylogeny: a study of the phenomena of neurobiotaxis. Part 1. Cyclostomi and Pisces. Journal of Comparative Neurology 27(4):467-564. (3,4,5)

50. Blatchley, W. S. 1938. The fishes of Indiana — with descriptions, notes on habits and distribution in the state. The Nature Publishing Company, Indianapolis, Indiana, USA. 121 pages. (1,6)

51. Bloch, M. E., and J. G. Schneider. 1801. Systema ichthyologiae inconibus ex illustratum. Berlin, Germany. 584 pages. (Facsimile report, Stechert, New York. 1967) (1,2,4)

52. Boehmer, R. J. 1973. Ages, lengths, and weights of paddlefish caught in Gavins Point Dam Tailwaters, Nebraska. Proceedings of the South Dakota Academy of Sciences 52:140-146. (13)

53. Bogue, R. 1982. James River paddlefish. Fins and Feathers (Missouri) 4(3):82-85. (6,7)

54. Bond, J. 1966. Spoonbill spectacular. Outdoor Oklahoma 22(6):12-13. (7,8)

55. Bonislawsky, P. S. 1977. Paddlefish investigation. Kansas Fish and Game Commission. Dingell-Johnson Project F-15-R, Study 030. Final Report. 18 pages. (8,10,13)

56. Bonn, E. W., and C. R. Inman. 1955. Basic survey and inventory of species present and their distribution in the Sulphur River in Texas. Texas Parks and Wildlife Department. Dingell-Johnson Project F-8-R-2, Job Number 1. Progress Report. 20 pages. (1)

57. Bourtzev, I. A., and A. D. Gershanovich. 1975. Paddlefish success in Russia. Aquaculture and the Fish Farmer 2(1):28. (14)

58. Bowers, C. C., Jr. 1980. Special permit gill and trammel net fishing, November 1, 1979 to March 31, 1980. Kentucky Department of Fish and Wildlife, Division of Fisheries, Frankfort, Kentucky, USA. 7 pages. (9)

59. Brady, D. R. 1980. Age determination utilizing calcified annuli in the dentary bone of the paddlefish (*Polyodon spathula*). Proceedings of the Nebraska Academy of Sciences and Affiliated Societies 90:5. (abstract of a presentation) (13)

60. Brady, D. R. 1981. Caudal neurosecretory system of the paddlefish, *Polyodon spathula* (Osteichthyes: Chondrostei): a light microscopic study. Master of Arts Thesis. University of Nebraska at Omaha, Omaha, Nebraska, USA. 39 pages. (5)

61. Brady, D. R. 1984. Caudal neurosecretory system of the paddlefish, *Polyodon spathula*. Zoologischer Anzeiger 213(3-4):215-223. (5)

62. Brady, D. R., and T. A. Georgi. 1978. Morphological and histological evaluation of the *Polyodon spathula* oral cavity. Proceedings of the Nebraska Academy of Sciences and Affiliated Societies. 88:7-8. (abstract of a presentation) (4)

63. Brandt, R. L. 1978. Fungal control methods, diets and water temperatures used to culture paddlefish, *Polyodon spathula*. Master of Science Thesis.

South Dakota State University, Brookings, South Dakota, USA. 39 pages. (12,13,14)

64. BRANSON, B. A. 1974. American paddlefish: signs of distress. National Parks and Conservation Magazine 48(1):21-23. (6,7)

65. BRANSON, B. A. 1977A. Endangered fish of Kentucky streams. Natural History 86(2):64-69. (6,7)

66. BRANSON, B. A. 1977B. Threatened fishes of Daniel Boone National Forest, Kentucky, USA. Transactions of the Kentucky Academy of Science 38(1-2):69-73. (1,6)

67. BRIDGE, T. W. 1878. On the osteology of *Polyodon folium*. Philosophical Transactions of the Royal Society of London 169 (part 2):683-733. (4)

68. BRIDGE, T. W. 1896. The mesial fins of ganoids and teleosts. Journal of the Linnean Society of London, Zoology 25:530-602. (4)

69. BRIDGE, T. W. 1897A. On the presence of ribs in *Polyodon (Spatularia) folium*. Proceedings of the Zoological Society of London. 1897:722-724. (4)

70. BRIDGE, T. W. 1897B. *Polyodon* has cartilaginous ribs. Zoologischer Anzeiger 20:240. (4)

71. BRONTE, C. R., AND D. W. JOHNSON. 1984. Evaluation of commercial entanglement — gear fishery in Lake Barkley and Kentucky Lake, Kentucky. North American Journal of Fisheries Management 4(1):75-83. (6,9)

72. BRONTE, C. R., AND D. W. JOHNSON. 1985. Growth of paddlefish in two mainstream reservoirs with reference to commercial harvest. Transactions of the Kentucky Academy of Science 46(1-2):28-32. (6,9,10,13)

73. BROWN, C. J. D. 1951. The paddlefish in Fort Peck Reservoir, Montana. Copeia 1951(3):252. (1,6)

74. BROWN, C. J. D. 1962. A preliminary list of Montana fishes. Proceedings of the Montana Academy of Sciences 22:21-26. (1)

75. BROWN, C. J. D. 1971. Fishes of Montana. Big Sky Books, Montana State University, Bozeman, Montana, USA. 207 pages. (1,6,13)

76. BRYAN, P. 1942. Spoonbill fishing, a new valley industry. Alabama Conservationist 14(12):7, 15. (7,9)

77. BRYAN, P. 1945. Baby spoonbill caught. Alabama Conservationist. 16(7):7-12. (6)

78. BRYAN, P., AND C. M. TARZWELL. 1941. A preliminary report on the census of commercial fishing in TVA impoundments. Transactions of the North American Wildlife Conference. 6:256-272. (9)

79. BRYAN, P., AND C. E. WHITE, JR. 1958. An economic evaluation of the commercial fishing industry in the T.V.A. Lakes of Alabama during 1956. Proceedings of the Annual Conference Southeastern Association of Game and Fish Commissioners 12:128-132. (9)

80. BUCHANAN, T. M. 1973. Key to the fishes of Arkansas. Arkansas Game and Fish Commission, Little Rock, Arkansas, USA. 68 pages. (1)

81. BUCHANAN, T. M. 1974. Threatened native fishes of Arkansas. Page 79 in

Arkansas natural area plan. Arkansas Department of Planning, Little Rock, Arkansas, USA. 248 pages. (1,6)

82. Bumpas, G. S. 1973. Paddlefish cultivation possible. The American Fish Farmer and World Aquaculture News 4(6): 5-7. (14)

83. Burch, M. 1978. Spoonbills — big game fish of the Ozarks. Southern Outdoors 2:34-36. (6,7)

84. Burch, M. 1984. Fishing for spoonbills. Fins and Feathers (Missouri) 6(3):18-21. (6,7,8)

85. Burke, J. S. 1984. Influence of planktivorous fishes on zooplankton of catfish culture ponds. Master of Science Thesis. Auburn University, Auburn, Alabama, USA. 87 pages. (6,9,12,14)

86. Burr, B. M. 1980. *Polyodon spathula* (Walbaum), paddlefish. Pages 45-46 *in* D. S. Lee, C. R. Gilbert, C. H. Hocutt, R. E. Jenkins, D. E. McAllister, and J. R. Stauffer, Jr., editors. Atlas of North American Freshwater Fishes. North Carolina State Museum of Natural History, Raleigh, North Carolina, USA. 854 pages. (1,10)

87. Burtsev, I. A., and A. D. Gershanovich. 1976. On acclimatization of the paddlefish in the USSR. Rybnoe kh-vo, Number 10. Pages 13-16. (6,12,13)

88. Busch, R. 1974. Culture of paddlefish fry. Fisheries Research Annual Report. Auburn University, Auburn, Alabama, USA. 2:60-63. (14)

89. Busch, R., and E. W. Shell. 1973. Culture of paddlefish fry in a 1.4 acre earthen pond. Fisheries Research Annual Report. Auburn University, Auburn, Alabama, USA. 2:73-74. (14)

90. Cabot, S. 1848. A specimen of the shovel fish from the Ohio River. Proceedings of the Boston Society of Natural History 2:184-185. (6)

91. Call, R. E. 1896. Fishes and shells of the Falls of the Ohio. Memorial History of Louisville 1:9-20. (1)

92. Carlander, H. B. 1954. A history of fish and fishing in the Upper Mississippi River. Upper Mississippi River Conservation Committee, Rock Island, Illinois, USA. 96 pages. (6,9)

93. Carlander, K. D. 1969. Handbook of freshwater fishery biology. Volume 1. Iowa State University Press, Ames, Iowa, USA. 752 pages. (6,10,13)

94. Carlson, D. M. 1977. Epitaph for Lake Ozark paddlefish. Missouri Conservationist 38(4):10-13. (6,7)

95. Carlson, D. M., and P. S. Bonislawsky. 1981. The paddlefish (*Polyodon spathula*) fisheries of the midwestern United States. Fisheries 6(2):17-22, 26-27. (1,6,8,9,10)

96. Carlson, D. M., M. K. Kettler, S. E. Fisher, and G. S. Whitt. 1982. Low genetic variability in paddlefish populations. Copeia 1982 (3):721-725. (3,5)

97. Carroll, B. B., G. E. Hall, and R. D. Bishop. 1963. Three seasons of rough fish removal at Norris Reservoir, Tennessee. Transactions of the American Fisheries Society 92(4):356-364. (9)

98. Carter, B. T. 1954. The movement of fishes through navigational lock chambers in the Kentucky River. Transactions of the Kentucky Academy of Science 15(3):48-56. (1)

99. Carter, E. R. 1952. Kentucky Lake investigations. Rotenone samplings. Kentucky Department of Fish and Wildlife Resources. Dingell-Johnson Project F-2-R-1, Job D. Progress Report. 2 pages. (1)

100. Charles, J. R. 1964. Biological and limnological studies of streams. Effects of oil field brines on biological productivity of the Green River in Kentucky. Kentucky Department of Fish and Wildlife Resources. Dingell-Johnson Project F-18-R-4. Final Report. 82 pages. (1)

101. Churchill, E. P., and W. H. Over. 1933. Fishes of South Dakota. South Dakota Department of Game and Fish, Pierre, South Dakota, USA. 87 pages. (1,6)

102. Clawson, C. C., J. Finstad, and R. A. Good. 1966. Evolution of the immune response. V. Electron microscopy of plasma cells and lymphoid tissue of the paddlefish. Laboratory Investigations 15(12):1820-1847. (5)

103. Clay, W. M. 1975. The fishes of Kentucky. Kentucky Department of Fish and Wildlife Resources, Frankfort, Kentucky, USA. 416 pages. (1,6)

104. Coker, R. E. 1923. Methuselah of the Mississippi. Scientific Monthly 16(1):89-103. (6)

105. Coker, R. E. 1930a. Keokuk Dam and the fisheries of the upper Mississippi River. Bulletin of the United States Bureau of Fisheries 45(1929):87-139. (6,9)

106. Coker, R. E. 1930b. Studies of common fishes of the Mississippi River to Keokuk. Bulletin of the United States Bureau of Fisheries 45(1929):141-225. (6,9)

107. Collinge, W. E. 1894. The sensory canal system of fishes. Part 1. Ganoidei. Quarterly Journal of Microscopical Science 36(144 n.s.):499-537. (4)

108. Collinge, W. E. 1895. On the presence of scales in the integument of *Polyodon folium*. Journal of Analytical Physiology (Paris) 29:485-487. (4)

109. Combs, D. L. 1979. Paddlefish investigations: fish research and surveys for Oklahoma lakes and reservoirs. Oklahoma Department of Wildlife Conservation. Dingell-Johnson Project F-37-R-2. Progress Report. 11 pages. (8,9,10,13)

110. Combs, D. L. 1980. Paddlefish investigations: fish research and surveys for Oklahoma lakes and reservoirs. Oklahoma Department of Wildlife Conservation. Dingell-Johnson Project F-37-R-3. Progress Report. 11 pages. (8,9,10,13)

111. Combs, D. L. 1981. Paddlefish investigations: fish research and surveys for Oklahoma lakes and reservoirs. Oklahoma Department of Wildlife Conservation. Dingell-Johnson Project F-37-R-4. Final Report. 42 pages. (8,9,10,13)

112. Combs, D. L. 1982. Angler exploitation of paddlefish in the Neosho River, Oklahoma. North American Journal of Fisheries Management 2(4):334-342. (8,9,10,13)

113. Combs, D. L. 1984. Paddlefish investigations: fish research and surveys for Oklahoma lakes and reservoirs. Oklahoma Department of Wildlife Conservation. Dingell-Johnson Project F-37-R-6. Progress Report. 8 pages. (8,9,10,13)

114. COMBS, D. L. 1986. The role of regulations in managing paddlefish populations. Pages 68-76 *in* J. G. Dillard, L. K. Graham, and T. R. Russell, editors. The paddlefish: status, management and propagation. North Central Division, American Fisheries Society, Special Publication Number 7. 159 pages. (6,8,10)

115. CONNER, J. V. 1977. Zoogeography of the fishes in the Gulf Slope Drainages: Calcasieu to the Rio Grande. Doctoral Dissertation. Tulane University, New Orleans, Louisiana, USA. 266 pages. (1)

116. COOK, F. A. 1958. Early history and trends in Mississippi freshwater fisheries. Mississippi Game and Fish Commission. Jackson, Mississippi, USA. 35 pages. (9)

117. COOK, F. A. 1959. Freshwater fishes in Mississippi. Mississippi Game and Fish Commission, Jackson, Mississippi, USA. 239 pages. (1,6)

118. COPE, E. D. 1883. A new chondrostean from the Eocene. American Naturalist 17:1152-1153. (3)

119. COPE, E. D. 1885. Eocene paddlefish and Gonorhynchidae. American Naturalist 19:1090-1091. (3)

120. COPE, E. D. 1886. On two new forms of Polyodont and Gonorhynchid fishes from the Eocene of the Rocky Mountains. Memoirs of the National Academy of Science 3(part 2):161-165. (3)

121. COTTREL, T. 1944. Rough fish given hook treatment — Big Lake clubmen reduce the population of spoonbills, gars, and turtles. Alabama Conservationist 15(7):4. (7)

122. COWAN, C. M. 1966. Protein profiles in lower fish. Proceedings of the Nebraska Academy of Sciences 76:5-6. (abstract of a presentation) (5)

123. CRANCE, J. H. 1985. Paddlefish riverine habitat suitability index curves developed by the delphi technique for use with instream flow incremental methodology. United States Fish and Wildlife Service, Western Energy and Land Use Team, Fort Collins, Colorado, USA. 36 pages. (6)

124. CROSS, F. B. 1967. Handbook of fishes of Kansas. University of Kansas Museum of Natural History Miscellaneous Publication Number 45. Lawrence, Kansas, USA. 357 pages. (1,6)

125. CVANCARA, V. A. 1974. Liver carabamyl phosphate synthetase in the primitive freshwater bony fishes (Chondrostei, Holostei). Comparative Biochemistry and Physiology 49b(4):785-787. (5)

126. DANFORTH, C. H. 1911. A 74 millimeter *Polyodon*. Biological Bulletin (Woods Hole). 20(4):201-204. (6,11)

127. DANFORTH, C. H. 1912. The heart and arteries of *Polyodon*. Journal of Morphology 23(3):409-454. (4)

128. DANFORTH, C. H. 1913. The myology of *Polyodon*. Journal of Morphology 24(1):107-146. (4)

129. DAVIS, P. W., JR. 1959. The yellow bodies of the chondrostean fish, *Polyodon spathula*. Bios 30(3):155-158. (5)

130. DEACON, J. E., G. KOBETICH, J. D. WILLIAMS, AND S. CONTERAS, AND OTHER

MEMBERS OF THE ENDANGERED SPECIES COMMITTEE OF THE AMERICAN FISHERIES SOCIETY. 1979. Fishes of North America endangered, threatened or of special concern: 1979. Fisheries 4(2):29-41. (6)

131. DEAN, B. 1895. Fishes, living and fossil. Columbia University Biological Series 3. Macmillan and Company, New York, New York, USA. 300 pages. (3)

132. DEETHARDT, D. 1977. Paddlefish. South Dakota Agricultural Experimental Station Bulletin Number 651, Brookings, South Dakota, USA. 18 pages. (pamphlet on paddlefish recipes) (7)

133. DEWEY, M. R., AND T. E. MOEN. 1978. Fishes of the Caddo River, Arkansas after impoundment of De Gray Lake. Arkansas Academy of Science 32:39-42. (1)

134. DINGERKUS, G., AND W. M. HOWELL. 1976. Karyotypic analysis and evidence of tetraploidy in the North American paddlefish, *Polyodon spathula.* Science 194(4267):842-844. (5)

135. DOROSHOV, S. I. 1977. The present status and perspectives for artificial rearing and acclimation of sea and brackish water fishes in the USSR. Proceedings of the Third Japan-Soviet Joint Symposium on Aquaculture. November 1974, Tokyo, Japan. Pages 63-73. (14)

136. DOUGLAS, N. H. 1974. Freshwater fishes of Louisiana. Claitor's Publishing Division, Baton Rouge, Louisiana, USA. 443 pages. (1,6)

137. DOWNEY, H. 1909. The lymphatic tissue of the kidney of *Polyodon spathula.* Folia Haematologica (Leipzig). 8:415-466. (4)

138. DOWNEY, H. 1910. Phagocytosis of erythrocytes in the lymphorenal tissue of *Polyodon spathula.* Folia Haematologica. (Leipzig). 9:81-86. (5)

139. DOWNEY, H. 1911. The origin and structure of the plasma cells of normal vertebrates, especially of the cold-blooded vertebrates, and the eosinophils of the lung amblystoma (plasma cells in *Polyodon spathula*). Folia Haematologica (Leipzig) 11:275-314. (5)

140. DRAGOMIROV, N. I., AND M. P. NIKOL'SKAYA. 1978. Gradients in the development of the spiracle-branchial row of neuromasts during ontogenesis of sturgeon. Doklady Biological Sciences, Doklady Akademii Nauk SSSR, Seriya Biologiya 241(1-6):332-334. (4,5)

141. DUMERIL, A. H. 1868. Note sur trois poissons: *Acipenser dabryanus, Polyodon gladius,* et *Peristhethidion prionocephalum,* accompagnee de quelques considerations generales sur les groupes auxquels ces especes appartiennent. Nouvelles des Archives du Musee d'Histoire Naturelle Paris 4:93-116. (16)

142. EDDY, S. 1945. Little-known fishes of Minnesota. Minnesota Department of Conservation, Conservation Bulletin Number 9, St. Paul, Minnesota, USA. Pages 21-24. (1,6)

143. EDDY, S., AND P. SIMER. 1929. Notes on the food of the paddlefish and the plankton of its habitat. Transactions of the Illinois State Academy of Science 25:59-68. (12)

144. EDDY, S., AND J. C. UNDERHILL. 1974. Northern fishes, 3rd edition. University

of Minnesota Press, Minneapolis, Minnesota, USA. 414 pages. (1,6)

145. ELKIN, R. E., JR. 1959. Commercial fisheries catch in Oklahoma, 1957. Proceedings of the Oklahoma Academy of Science (1958) 39:183-190. (9)

146. ELLIS, M. M. 1916. *Polyodon* in western Missouri. Copeia 1916:93-94. (1)

147. ELSER, A. A. 1972. Fish management surveys. Montana Department of Fish and Game. Dingell-Johnson Project F-30-R-7, Job Number 1-B. Progress Report. 8 pages. (6,8,10)

148. ELSER, A. A. 1973A. The great paddlefish puzzle. Montana Outdoors 4(3):30-34. (6,7,8)

149. ELSER, A. A. 1973B. Fish management surveys. Montana Department of Fish and Game. Dingell-Johnson Project F-30-R-8, Job Number 1-B. Progress Report. 15 pages. (6,8,10)

150. ELSER, A. A. 1975. Paddlefish investigations. Montana Department of Fish and Game. Dingell-Johnson Project F-30-R-11, Job Number 2-A. Progress Report. 11 pages. (6,8,10)

151. ELSER, A. A. 1976. Paddlefish investigations. Montana Department of Fish and Game. Dingell-Johnson Project F-30-R-12, Job Number 2-A. Progress Report. 12 pages. (6,8,10)

152. ELSER, A. A. 1977. Paddlefish investigations. Montana Department of Fish and Game. Dingell-Johnson Project F-30-R-13, Job Number 2-A. Progress Report. 14 pages. (6,8,10)

153. ELSER, A. A. 1978. Paddlefish investigations. Montana Department of Fish, Wildlife, and Parks. Dingell-Johnson Project F-30-R-14, Job Number 2-A. Progress Report. 16 pages. (6,8,10)

154. ELSER, A. A. 1979. Paddlefish investigations. Montana Department of Fish, Wildlife, and Parks. Dingell-Johnson Project F-30-R-15, Job Number 2-A. Progress Report. 16 pages. (6,8,10)

155. ELSER, A. A. 1980. Paddlefish investigations. Montana Department of Fish, Wildlife, and Parks. Dingell-Johnson Project F-30-R-16, Job Number 2-A. Progress Report. 15 pages. (6,8,10)

156. ELSER, A. A. 1981. Paddlefish investigations. Montana Department of Fish, Wildlife, and Parks. Dingell-Johnson Project F-30-R-17, Job Number 2-A. Progress Report. 14 pages. (6,8,10)

157. ELSER, A. A. 1982. Paddlefish investigations. Montana Department of Fish, Wildlife, and Parks. Dingell-Johnson Project F-30-R-18, Job Number 2-A. Progress Report. 15 pages. (6,8,10)

158. ELSER, A. A. 1983. Paddlefish investigations. Montana Department of Fish, Wildlife, and Parks. Dingell-Johnson Project F-30-R-19, Job Number 2-A. Progress Report. 14 pages. (6,8,10)

159. ELSER, A. A. 1986. An overview of current management practices for paddlefish fisheries. Pages 62-67 *in* J. G. Dillard, L. K. Graham, and T. R. Russell, editors. The paddlefish: status, management and propagation. North Central Division, American Fisheries Society, Special Publication Number 7. 159 pages. (6,8,10)

160. Estes, R. D., and T. B. Hess. 1974. Fisheries survey on Old Hickory Reservoir near Dixon Springs, Tennessee. Annual Report, September 1973-September 1974. Tennessee Valley Authority, Division of Forestry, Fisheries and Wildlife, Norris, Tennessee, USA. 62 pages. (1,11)

161. Evermann, B. W. 1902. Description of a new species of shad (*Alosa ohiensis*) with notes on other food fishes of the Ohio River. Report of the United States Commission on Fish and Fisheries for the year ending June 30, 1901. Pages 273-288. (9)

162. Evermann, B. W. 1918. The fishes of Kentucky and Tennessee: a distributional catalogue of the known species. Bulletin, U.S. Bureau of Fisheries, Document 858:293-368. (1)

163. Evermann, B. W., and U. O. Cox, 1896. A report upon the fishes of the Missouri River Basin. Report of the United States Commission on Fish and Fisheries for the year ending June 30, 1894. Pages 325-429. (1)

164. Evermann, B. W., and E. L. Goldsborough. 1902. Notes on the fishes and mollusks of Lake Chautauqua, New York. Report of the United States Commission on Fish and Fisheries for the year ending June 30, 1901. Pages 169-175. (1,6)

165. Faxon, W. 1915. Relics of Peal's Museum. Bulletin of the Museum of Comparative Zoology, Harvard University. 59(3):119-150. (1,2,4)

166. Fiedler, R. H. 1933. Fisheries industries of the United States, 1932. Report of the United States Commissioners of Fisheries. Appendix 3. Pages 149-449. (9)

167. Finstad, J., and R. A. Good. 1966. Phylogenetic studies of adaptive immune responses in the lower invertebrates. Pages 173-188 *in* R. T. Smith, P. A. Miescher, and R. A. Good, editors. Phylogeny of immunity. University of Florida Press, Gainesville, Florida, USA. 276 pages. (5)

168. Fischer, H. 1978. Behold the paddlefish. Defender's 53(1):24-26. (7)

169. Fish, L. A., B. Pollara, and R. A. Good. 1966. Characterization of an immunoglobulin from the paddlefish (*Polyodon spathula*). Pages 99-104 *in* R. T. Smith, P. A. Miescher, and R. A. Good, editors. Phylogeny of immunity. University of Florida Press, Gainesville, Florida, USA. 276 pages. (5)

170. Fisher, H. J. 1962. Some fishes of the lower Missouri River. American Midland Naturalist 68(2):424-429. (1)

171. Fisheries Service Bulletin. 1919. Rearing paddlefish in confinement. United States Bureau of Fisheries, Fisheries Service Bulletin Number 48:2. (14)

172. Fitz, R. B. 1966. Unusual food of a paddlefish (*Polyodon spathula*) in Tennessee. Copeia 1966(2):356. (12)

173. Flury, A. 1965. Super snooper. Texas Parks and Wildlife Magazine 23(9):10. (7)

174. Fogle, N. E. 1961. Statewide fisheries investigations during the fourth year of impoundment of Oahe Reservoir. South Dakota Department of Game,

Fish, and Parks. Dingell-Johnson Project F-1-R-11, Job Numbers 9-12. Progress Report. 47 pages. (8,10)

175. Fogle, N. E. 1963. Report of fisheries investigations during the fourth year of impoundment of Oahe Reservoir, South Dakota, 1961. South Dakota Department of Game, Fish, and Parks. Dingell-Johnson Project F-1-R-11, Job Numbers 10-12. Progress Report. 36 pages. (8,10)

176. Forbes, S. A. 1878. The food of Illinois fishes. Illinois State Laboratory of Natural History Bulletin 1(2):71-89. (12)

177. Forbes, S. A. 1884. A catalogue of the native fishes of Illinois. Report of the Illinois State Fish Commissioner for 1884. Springfield, Illinois, USA. Pages 60-89. (1)

178. Forbes, S. A. 1888a. Studies of the food of freshwater fishes. Illinois State Laboratory of Natural History Bulletin 2:433-473. (12)

179. Forbes, S. A. 1888b. On the food relations of freshwater fishes; a summary and discussion. Illinois State Laboratory of Natural History Bulletin 2:475-538. (12)

180. Forbes, S. A., and R. E. Richardson. 1908. The fishes of Illinois. Illinois State Laboratory of Natural History, Urbana, Illinois, USA. 357 pages. (contains a separate atlas with 102 maps) (1)

181. Forbes, S. A., and R. E. Richardson. 1920. The fishes of Illinois, 2nd edition. Illinois Natural History Survey, Urbana, Illinois, USA. 357 pages. (1,6,9)

182. Fowler, H. W. 1919. A list of the fishes of Pennsylvania. Biennial Report. Proceedings of the Biological Society of Washington 32:49-74. (1)

183. Fowler, H. W. 1921. Description of a new cyprinoid fish (*Notropis stonei*) with notes on other fishes obtained in the United States. Proceedings of The Academy of Natural Sciences of Philadelphia 72:385-402. (1)

184. Fowler, H. W. 1940. A list of the fishes recorded from Pennsylvania. Biennial Report. Pennsylvania Board of Fish Commissioners, Harrisburg, Pennsylvania, USA. 25 pages. (1)

185. Friberg, D. V. 1972a. Living relics. South Dakota Conservation Digest 39:28-29. (7,8)

186. Friberg, D. V. 1972b. Paddlefish abundance and harvest within a population lacking recruitment, Big Bend Dam Tailwaters, 1961-1971, South Dakota. South Dakota Department of Game, Fish, and Parks. National Marine Fisheries Service Project 4-61-R, Segment 1-3. Completion Report. 16 pages. (6,8,10,11,13)

187. Friberg, D. V. 1973. Investigation of paddlefish populations in South Dakota and the development of management plans, 1972. South Dakota Department of Game, Fish, and Parks. Dingell-Johnson Project F-15-R-7, Study 9, Job Numbers 1-5. Progress Report. 37 pages. (6,8,10,13,14)

188. Friberg, D. V. 1974. Investigation of paddlefish populations in South Dakota and development of management plans, 1973. South Dakota Department of Game, Fish, and Parks. Dingell-Johnson Project F-15-R-8, Study 9, Job Numbers 1-5. Progress Report. 33 pages. (6,8,10,13,14)

189. FRY, J. P. 1962. Harvest of fish from tailwaters of three large impoundments in Missouri. Proceedings of the Annual Conference Southeastern Association of Game and Fish Commissioners 16:405-411. (8)

190. GANNON, J. E., AND R. P. HOWMILLER. 1973. Ecological notes on paddlefish (*Polyodon spathula*) with short rostrums. Michigan Academician 6(2):217-222. (4)

191. GARDNER, W. M., AND R. K. BERG. 1982. An analysis of the instream flow requirements for selected fishes in the wild and scenic portion of the Missouri River. Montana Department of Fish, Wildlife, and Parks, Great Falls, Montana, USA. 111 pages. (1,6,10)

192. GARMAN, H. 1890. A preliminary report on the animals of the Mississippi bottoms near Quincy, Illinois, in August, 1888. Part 1. Illinois State Laboratory of Natural History Bulletin 3:123-184. (1,6)

193. GARMAN, H. 1896. Some notes on the brain and pineal structures of *Polyodon folium*. Illinois State Laboratory of Natural History Bulletin 4:298-309. (4)

194. GENGERKE, T. W. 1977. Paddlefish investigations. Iowa Conservation Commission. United States National Marine Fisheries Service Project 2-255-R, Segment 2. Progress Report. 15 pages. (6,8,9,10,11,13)

195. GENGERKE, T. W. 1978. Paddlefish investigations. Iowa Conservation Commission. United States National Marine Fisheries Service Project 2-225-R, Segment 1-3. Final Report. 86 pages. (8,9,10,13)

196. GENGERKE, T. W. 1986. Distribution and abundance of paddlefish in the United States. Pages 22-35 *in* J. G. Dillard, L. K. Graham, and T. R. Russell, editors. The paddlefish: status, management and propagation. North Central Division, American Fisheries Society, Special Publication Number 7. 159 pages.(1,8,9)

197. GEORGI, T. A. 1981. A preliminary study of the electroreceptive capability of the paddlefish *Polyodon spathula* (Walbaum). Proceedings of the Nebraska Academy of Sciences and Affiliated Societies 91:9. (abstract of a presentation) (5)

198. GEORGI, T. A., AND D. BEEDLE. 1978. The histology of the excretory kidney of the paddlefish (*Polyodon spathula*). Journal of Fisheries Biology 12:587-590. (4,5)

199. GERKING, S. D. 1945. The distribution of the fishes of Indiana. Investigations of Indiana Lakes and Streams 3(1):1-137. (1,6)

200. GERSHANOVICH, A. D. 1979. Respiratory metabolism of paddlefish juveniles in grow-out systems. Pages 74-75 *in* V. I. Lukyanenko, editor. Ecological physiology and biochemistry of fish. 4th All-Union Conference, Volume No. 2, September 1979, Askrakham, USSR. (5,14)

201. GERSHANOVICH, A. D. 1980. Energy metabolism and food requirements of young paddlefish. Pages 27-32 *in* Fiziologiva morskikh rybnoe (The physiology of sea fishes) Pishchevaya promyshlernost (Moscow). (5,6,12,13)

202. GERSHANOVICH, A. D. 1983. Factors determining variations in growth rate and size distribution in groups of young paddlefish, *Polyodon spathula* (Polyodontidae). Journal of Ichthyology 23(4):56-61. (6,10,13)

203. Gershanovich, A. D., and A. I. Nikolaev. 1985. Results of rearing paddlefish in polyculture with sturgeon hybrids, carp and herbivorous fish within 3 years. Pages 170-175 *in* Marine fish farming. Collected papers. All-Union Research Institute of Marine Fisheries and Oceanography (Moscow). (6,10,13,14)

204. Gilbert, B. 1981. End of a long journey for spoonbill cat. Audubon 83(2):66-71. (6,7)

205. Good, R. A., and J. Finstad. 1964. Phylogeny of immune reactions: immunological capacity of *Polyodon* and other fishes. Federal Proceedings, Federation of American Societies for Experimental Biology 23:285. (5)

206. Good, R. A., J. Finstad, B. Pollara, and A. E. Gabrielsen. 1966. Pages 149-168 *in* R. T. Smith, P. A. Miescher, and R. A. Good, editors. Phylogeny of immunity. University of Florida Press, Gainesville, Florida, USA. 276 pages. (5)

207. Good, R. A., A. E. Gabrielsen, B. Pollara, H. Gewurz, J. Finstad, and B. Cinader. 1971. Regulation of antibody response. 2nd edition. C. C. Thomas. Springfield, Illinois, USA. 400 pages. (5)

208. Good, R. A., R. D. A. Peterson, J. Finstad, and A. E. Gabrielsen. 1965. Morphologic studies of the development of the lymphoid tissues. Series Haematologica (1965) 8:1-28. (5)

209. Gowanloch, J. N. 1933. Fishes and fishing in Louisiana. Louisiana Department of Conservation Bulletin Number 23, Baton Rouge, Louisiana, USA. 638 pages. (1,6)

210. Graham, L. K. 1980a. Maintenance of the Osage River paddlefish fishery. Missouri Department of Conservation. Dingell-Johnson Project F-1-R-29, Study Number S-28. Progress Report. 4 pages. (6,8,10)

211. Graham, L. K. 1980b. Paddlefish: Good news and bad news. Missouri Conservationist 41(3):20-22. (6,7)

212. Graham, L. K. 1981. Maintenance of the Osage River paddlefish fishery. Missouri Department of Conservation. Dingell-Johnson Project F-1-R-30, Study Number S-28. Progress Report. 4 pages. (6,8,10)

213. Graham, L. K. 1982. Maintenance of the Osage River paddlefish fishery. Missouri Department of Conservation. Dingell-Johnson Project F-1-R-31, Study Number S-28. Progress Report. 4 pages. (6,8,10)

214. Graham, L. K. 1983. Maintenance of the Osage River paddlefish fishery. Missouri Department of Conservation. Dingell-Johnson Project F-1-R-32, Study Number S-28. Progress Report. 4 pages. (6,8,10)

215. Graham, L. K. 1984a. Management plan for paddlefish in Missouri (in preparation). Missouri Department of Conservation, Columbia, Missouri, USA. (6,8,10,11,13)

216. Graham, L. K. 1984b. Maintenance of the Osage River paddlefish fishery. Missouri Department of Conservation. Dingell-Johnson Project F-1-R-33, Study Number S-28. Progress Report. 6 pages. (6,8,10)

217. GRAHAM, L. K. 1985. Maintenance of the Osage River paddlefish fishery. Missouri Department of Conservation. Dingell-Johnson Project F-1-R-34, Study Number S-28. Progress Report. 6 pages. (6,8,10)

218. GRAHAM, L. K. 1986. Establishing and maintaining paddlefish populations by stocking. Pages 96-104 *in* J. G. Dillard, L. K. Graham, and T. R. Russell, editors. The paddlefish: status, management and propagation. North Central Division, American Fisheries Society, Special Publication Number 7. 159 pages. (6,8,10,11,13)

219. GRAHAM, L. K., AND P. S. BONISLAWSKY. 1978. An indexed bibliography of the paddlefish (*Polyodon spathula*). Missouri Department of Conservation, Columbia, Missouri, USA. 14 pages. (1,2,3,4,5,6,7,8,9,10,11,12,13,14,15,16)

220. GRAHAM, L. K., E. J. HAMILTON, T. R. RUSSELL, AND C. E. HICKS. 1986. The culture of paddlefish — a review of methods. Pages 78-94 *in* J. G. Dillard, L. K. Graham, and T. R. Russell, editors. The paddlefish: status, management and propagation. North Central Division, American Fisheries Society, Special Publication Number 7. 159 pages. (14)

221. GRANT, B. F., P. M. MEHRLE, AND T. R. RUSSELL. 1970. Serum characteristics of spawning paddlefish (*Polyodon spathula*). Comparative Biochemistry and Physiology 37:321-330. (5)

222. GRANTHAM, B. J. 1960. Detection of sources of stream pollution and its effects upon bottom organisms, fishes, and chemical qualities of the water. Completion report of pollution studies on the Pearl River. Mississippi Game and Fish Commission. Dingell-Johnson Project F-9-R-1. Final Report. 81 pages. (1)

223. GREEN, L. M., J. S. TUTEN, AND J. W. AVAULT, JR. 1978. Polyculture of Red Swamp Crayfish (*Procambarus clarkii*) and several North American fish species. Proceedings of the Fourth International Association of Astacology Symposium. Thonen-Les-Bains, France, August 28-31, 1978. 11 pages. (14)

224. GREEN, R. 1976. Handbook of Gentry County fishes. Albany Ledger, Albany, Missouri, USA. 41 pages. (1,7)

225. GREENE, C. W. 1935. The distribution of Wisconsin fishes. Wisconsin Conservation Commission, Madison, Wisconsin, USA. 235 pages. (1,6)

226. GREGORY, W. K. 1933. Fish skulls: a study of the evolution of natural mechanisms. Transactions of the American Philosophical Society (new series Number 2) 23:7-481. (3,4)

227. GROEN, C. L. 1973. A creel census survey of the Missouri River sport fishery. Master of Arts Thesis. University of South Dakota, Vermillion, South Dakota, USA. 126 pages. (8,10)

228. GROEN, C. L., AND J. C. SCHMULBACH. 1978. The sport fishery of the unchannelized and channelized Middle Missouri River. Transactions of the American Fisheries Society 107(3):412-418. (8)

229. GUNTHER, A. 1870. Catalogue of the fishes in the British Museum. Volume 8. British Museum, London, England. 549 pages. (1,2,4)

230. GUNTHER, A. 1873. Report on a collection of fishes from China. Annual Magazine of Natural History 12(series 4):239-250. (16)

231. Gunter, G. 1942. A list of fishes of the mainland of North and Middle America recorded from both freshwater and sea water. The American Midland Naturalist 28(2):305-326. (1,6)

232. Halkett, A. 1913. Checklist of the fishes of the Dominion of Canada and Newfoundland. King's Printer, Ottawa, Canada. 138 pages. (1)

233. Handyside, P. D. 1875. On the anatomy of a new species of *Polyodon*, the *Polyodon gladius* of Martens, taken from the river Yangtsze-Kiang, 450 miles above Woosung. Part 1. Its external characters and structure. Part 2. Its nervous and muscular systems. Proceedings of the Royal Society of Edinburgh 8:50-51, 136-137. (16)

234. Hansen, G. N. 1971. On the structure and vascularization of the pituitary gland in some primitive actinopterygians (*Acipenser, Polyodon, Calamoichthys, Polypterus, Lepisosteus,* and *Amia*). Kongelige Danske Videnskabernes Selskab Biologiske Skrifter 18(1):5-64. (4)

235. Hansen, G. N. 1983. Cell types in the adenohypophysis of the primitive actinopterygians, with special reference to immunocytochemical identification of pituitary hormone producing cells in the distal lobe. Pages 33-44 *in* Acta zoologica (1983 supplement). Royal Swedish Academy of Sciences, Pergamon Press, Elmsford, New York, USA. (5)

236. Hanson, W. D. 1976. The tailwater fisheries of Lake of the Ozarks and Pomme de Terre Lake, Missouri. Missouri Department of Conservation. Dingell-Johnson Project F-1-R-25, Study I-8, Job Number 1. Final Report. 18 pages. (8)

237. Hanson, W. D., and R. S. Campbell. 1963. The effects of pool size and beaver activity on distribution and abundance of warm-water fishes in a north Missouri stream. American Midland Naturalist 69(1):136-149. (1)

238. Harlan, J. R., and E. B. Speaker. 1969. Iowa fish and fishing, 4th edition. Iowa Conservation Commission, Des Moines, Iowa, USA. 377 pages. (1,6)

239. Harned, C. N. 1979. Management recommendations for *Polyodon spathula* (Walbaum) in Tennessee waters. Tennessee Valley Authority, Office of Natural Resources, Division of Air and Water Resources, Knoxville, Tennessee, USA. Progress Report. 7 pages. (6,9,10)

240. Harrow, L. G., and A. B. Schlesinger. 1980. The larval fish recruitment study. Omaha Public Power District, Environmental Series Bulletin Number 5, Omaha, Nebraska, USA. 92 pages. (11)

241. Hayashida, T. 1971. Biological and immunochemical studies with growth hormone in pituitary extracts of holostean and chondrostean fishes. General and Comparative Endocrinology 17(2):275-280. (5)

242. Hayashida, T. 1973. Biological and immunochemical studies with growth hormone in pituitary extracts of elasmobranchs. General and Comparative Endocrinology 20(2):377-385. (5)

243. Hayashida, T., and M. D. Lagios. 1969. Fish growth hormone: a biological, immunochemical, and ultrastructural study of sturgeon and paddlefish pituitaries. General and Comparative Endocrinology 13(3):403-411. (4,5)

244. Helms, D. R. 1973. Progress report on the second year study of shovelnose

sturgeon in the Mississippi River. Iowa Conservation Commission. National Marine Fisheries Service Project 2-156-R, Segment 2. Progress Report. 33 pages. (reported paddlefish fry in larval fish samples) (6,11)

245. HELMS, D. R. 1976. Paddlefish investigations. Iowa Conservation Commission. United States National Marine Fisheries Service Project 2-255-R, Segment 1. Progress Report. 18 pages. (6,8,9,10,11,13)

246. HEVEL, K. W., AND C. M. ALEXANDER. 1983. Trawling methodologies for juvenile paddlefish. Tennessee Valley Authority, Office of Natural Resources, Division of Air and Water Resources, Knoxville, Tennessee, USA. Progress Report. 18 pages. (6,10,12)

247. HILDEBRAND, S. F., AND I. L. TOWERS. 1927. Annotated list of fishes collected in the vicinity of Greenwood, Mississippi with descriptions of three new species. Bulletin of the United States Bureau of Fisheries 43 (part 2):105-136. (1)

248. HOAGUE, J. J., JR., R. WALLUS, AND L. K. KAY. 1976. Preliminary guide to the identification of larval fishes in the Tennessee River. Tennessee Valley Authority Technical Note B19. Knoxville, Tennessee, USA. 66 pages. (1,11)

249. HOLTON, G. 1980. The riddle of existence: fishes of "special concern". Montana Outdoors 11(1):2-6, 26. (6,7)

250. HOOGENBOON, K. J. H. 1928. Das Gehirn von *Polyodon folium*, Lacepède. Zeithschrift fuer Mikroskopisch-Anatomische Forschung. 18:311-392. (1,2,4)

251. HOOPES, D. T. 1960. Utilization of mayflies and caddis flies by some Mississippi River fishes. Transactions of the American Fisheries Society 89(1):32-34. (12)

252. HOUSER, A. 1965. Growth of paddlefish in Fort Gibson Reservoir, Oklahoma. Transactions of the American Fisheries Society 94(1):91-93. (13)

253. HOUSER, A., AND M. G. BROSS. 1959. Observations on the growth and reproduction of the paddlefish. Transactions of the American Fisheries Society 88(1):50-52. (11,13)

254. HOYT, R. D. 1984. Notes on various growth features of the paddlefish in the Ohio River. Transactions of the Kentucky Academy of Science 45(1-2):75-76. (6,13)

255. HUBBS, C. L. 1926. A checklist of the fishes of the Great Lakes and tributary waters, with nomenclatorial notes and analytical keys. University of Michigan, Museum of Zoology Miscellaneous Publication Number 15. 77 pages. (1)

256. HUBBS, C. L., AND K. F. LAGLER. 1964. Fishes of the Great Lakes Region. University of Michigan Press, Ann Arbor, Michigan, USA. 213 pages. (1,6)

257. HUBBS, C. 1976. A checklist of Texas freshwater fishes. Technical Series Number 11. Texas Parks and Wildlife Department, Austin, Texas, USA. 12 pages. (1)

258. HUBERT, W. A., S. H. ANDERSON, P. D. SOUTHALL, AND J. CRANCE. 1984. Habitat suitability index models and instream flow suitability curves:

paddlefish. United States Fish and Wildlife Service. FWS/DBS-82/10.80. 32 pages. (1,6,10)

259. HUSSAKOF, L. 1910. The spoonbill fishery of the lower Mississippi. Transactions of the American Fisheries Society 40:245-248. (6,9)

260. HUTCHINS, B. 1976. Rare paddlefish taken. Ohio Fisherman 2(3):28-29. (1,7)

261. IMMS, A. D. 1904. Notes on the gill rakers of the spoonbill sturgeon, *Polyodon spathula*. Proceedings of the Zoological Society of London (part 2):22-35. (4)

262. JENKINS, G., T. FORSYTHE, J. CONDER, AND C. BRONTE. 1982. Size selectivity and effects on gamefish populations of entanglement gear in Lake Barkley, Energy Lake, and Bard's Lake. Pages 80-90 *in* T. Crawford, M. Freeze, and S. H. Newton, editors. Proceedings of the 11th Annual Meeting — Inland Commercial Fisheries Association, March 2-4, 1982, North Little Rock, Arkansas, USA. 137 pages. (6,9)

263. JENKINS, R. E., AND J. A. MUSICK. 1980. Freshwater and marine fishes. Pages 319-373 *in* D. W. Linzey, editor. Endangered and threatened plants and animals of Virginia. Virginia Polytechnic Institute and State University, Blacksburg, Virginia, USA. 665 pages. (1,6)

264. JENKINS, R. M. 1953. Growth histories of the principal fishes in Grand Lake (O'the Cherokees), Oklahoma, through thirteen years of impoundment. Oklahoma Fishery Research Laboratory Report Number 34. 87 pages. (9)

265. JENSEN, M., AND B. JENSEN. 1979. The paddlers are running. Garcia Fishing Magazine 18:71-74. (6,7)

266. JOHNSON, D. W. 1981. Evaluation of the entanglement gear fishery in Lake Barkley and Kentucky Lake in Kentucky. National Marine Fisheries Service Project Number 2368-R-1. Annual Report. Murray State University, Murray, Kentucky, USA. 42 pages. (9,10)

267. JONES, L. 1961. Summary of commercial fisheries catch in Oklahoma for the years 1958-1960. Oklahoma Fishery Research Laboratory Report Number 83. 9 pages. (9)

268. JORDAN, D. S. 1878. A catalog of the fishes of Illinois. Illinois State Laboratory of Natural History Bulletin. 1(2):37-70. (1,6)

269. JORDAN, D. S. 1923. The genera of fishes and a classification of fishes. Reprinted 1963, Stanford University Press, Stanford, California, USA. 800 pages. (2,3,16)

270. JORDAN, D. S. 1925. Fishes, revised edition. D. Appleton and Company, New York, New York, USA. 773 pages. (2,6)

271. JORDAN, D. S., AND S. W. BRAYTON. 1878. On the distribution of the fishes of the Allegheny region of South Carolina, Georgia and Tennessee with descriptions of new or little known species. Bulletin of the United States National Museum Number 12. 96 pages. (1,6)

272. JORDAN, D. S., AND B. W. EVERMANN. 1896. The fishes of North and Middle America. Bulletin of the United States National Museum Number 47. Parts 1-4, 3313 pages. Reprinted 1963, for the Smithsonian Institution by T. F. H. Publications, Incorporated, Jersey City, New Jersey, USA. (2)

273. JORDAN, D. S., AND B. W. EVERMANN. 1922. American food and game fishes. Doubleday, Page, and Company, New York, New York, USA. 572 pages. (1,6)

274. JORDAN, D. S., B. W. EVERMANN, AND H. W. CLARK. 1930. A check list of the fishes and fish-like vertebrates of North and Middle America north of the northern boundary of Venezuela and Colombia. United States Bureau of Fisheries Report 1928, part 2. 670 pages. Facsimile report, E. Lundberg. 1962. (1,6)

275. JORGENSON, J., A. FLOCK, AND J. WERSALL. 1972. The Lorenzinian ampullae of *Polyodon spathula.* Zeitschrift fuer Zellforschung und Mikroskopische Anatomie 130:362-377. (4)

276. KALLEMEYN, L. W. 1974. Investigation of paddlefish populations in South Dakota and development of management plans, 1974. South Dakota Department of Game, Fish, and Parks. Dingell-Johnson Project F-15-R-9, Study 9, Job Numbers 1-8. Progress Report. 11 pages. (6,8,10,11,14)

277. KALLEMEYN, L. W. 1975A. Investigation of paddlefish populations in South Dakota. South Dakota Department of Game, Fish, and Parks. Dingell-Johnson Project F-15-R-10, Study 9, Job Numbers 2-4 and 7. Progress Report. 17 pages. (6,8,10,11,14)

278. KALLEMEYN, L. W. 1975B. Investigation of paddlefish populations in South Dakota and development of management plans, 1975. South Dakota Department of Game, Fish, and Parks. Dingell-Johnson Project F-15-R-11, Study 9, Job Numbers 3 and 7. Progress Report. 13 pages. (6,8,10,11,14)

279. KALLEMEYN, L. W. 1975C. Paddlefish — may be a future after all. South Dakota Conservation Digest 42:12-15. (7,14)

280. KALLEMEYN, L. W., AND J. F. NOVOTNY. 1977. Fish and fish food organisms in various habitats of the Missouri River in South Dakota, Nebraska, and Iowa. United States Fish and Wildlife Service, Office of Biological Services. FWS/OBS-77/25. Columbia, Missouri, USA. 100 pages. (6,11)

281. KAUP, J. J. 1862. Eine neue Art von spatularia. Archiv fuer Naturgeschichte 28 (part 1):278-281. (16)

282. KIMURA, S. 1934. Description of the fishes collected from the Yangtse-Kiang, China, by the late Dr. K. Kishinouye and his party in 1927-1929. Journal of the Shanghai Science Institute (section 3) 1:11-247. (16)

283. KIRKENDALL, K. R. 1983. Polyculture of catfish, chinese carps, and paddlefish. Master of Science Thesis. Auburn University, Auburn, Alabama, USA. 26 pages. (9,14)

284. KIRSH, P. 1892. Notes on the streams and fishes of Clinton County, Kentucky, with a description of a new darter. Bulletin of the United States Fish Commission (1890) 10:289-292. (1)

285. KIRTLAND, J. P. 1842-1845. Descriptions of the fishes of the Ohio River and its tributaries (part). Boston Journal of Natural History 4:16-26; 5:21-32. (1,2,4)

286. KIRTLAND, J. P. 1851. *Polyodon folium.* Lacepède. Spoonbill paddlefish. The Family Visitor, Cleveland and Hudson, Ohio, USA. 2(32):249. December 16. (1,2,4)

287. Kistler, H. D. 1906. The primitive pores of *Polyodon spathula*. Journal of Comparative Neurology and Psychology 16:294-298. (4)

288. Kofoid, C. A. 1900. Notes on the natural history of *Polyodon*. Science (new series) 11(268):252. (6)

289. Kreyenberg, M., and P. Pappenheim. 1908. Ein Beitrag zur Kenntnis der Fische des Yangtze und seiner Zuflusse. Sitzungsberichte der Gesellschaft der Naturforschenden Freunden zu Berlin. Pages 95-109. (16)

290. Kruger, R. 1985. Lining up for paddlefish. Fins and Feathers (Missouri). 7(3):11-14. (7,8)

291. Kuhne, E. R. 1939. A guide to the fishes of Tennessee and the Mid-South. Tennessee Department of Conservation, Nashville, Tennessee, USA. 124 pages. (1,6)

292. Kuzenski, S. 1977. Ancient technique could boost productivity of United States fish farms. Sea Grant Today 7(7):5. (14)

293. Lacepède, B. G. E. 1797. Mémoire sur le polyodon feuille. Bulletin des Sciences, Societe Philomathique, Paris. 1(2):49. (2)

294. Lacepède, B. G. E. 1798-1803. Histoire naturelle des poissons, Plassan, Paris. Buffon's Histoire Naturelle. Volumes 32-36. (2,4)

295. Lachner, E. A. 1956. The changing fish fauna of the Upper Ohio Basin. Pages 64-78 *in* Man and the waters of the Upper Ohio Basin. Special Publication Number 1, Pymatuning Laboratory of Field Biology, Linesville, Pennsylvania, USA. (1)

296. Lagios, M. D. 1968. The pituitary gland of the paddlefish, *Polyodon spathula*. Copeia 1968 (2):401-404. (4)

297. Lagios, M. D. 1973. Follicle boundary cells in the adenohypophysis of the chondrostean and holostean fishes: an ultrastructural study of their relationship to the follicular lumen, to endocrine cells, and to the hypophysial cleft. General and Comparative Endocrinology 20(2):362-376. (4)

298. Lagler, K. F., J. E. Bardach, and R. R. Miller. 1962. Ichthyology. John Wiley and Sons, New York, New York, USA. 545 pages. (6)

299. Large, T. 1903. A list of the native fishes of Illinois, with keys. Appendix to Report of the State Board of Fish Commissioners for September 30, 1900 to October 1, 1902. 30 pages. (1)

300. Larimore, R. W. 1949. Changes in the cranial nerves of the paddlefish, *Polyodon spathula,* accompanying development of the rostrum. Copeia 1949 (3):204-212. (4,5)

301. Larimore, R. W. 1950. Gametogenesis of *Polyodon spathula* (Walbaum): a basis for regulation of the fishery. Copeia 1950(2):116-124. (5,6,9,10,11,13)

302. Latimer, H. B. 1919. The lateral line of *Polyodon spathula*. Transactions of the American Microscopical Society 38(3):189-207. (4)

303. Legler, D. W., P. F. Weinheimer, R. T. Acton, H. K. Dupree, and T. R. Russell. 1971. Humoral immune factors in the paddlefish, *Polyodon spathula*. Comparative Biochemistry and Physiology 38B:523-527. (5)

304. LePage du Pratz, A. S. 1947. The history of Louisiana, or of the western parts of Virginia and Carolina. Notes by S. C. Arthur. Illustrations from the original French edition of 1758. J. S. W. Harmanson, New Orleans, Louisiana, USA 376 pages. (1,2,4)

305. LeSeur, C. A. 1818. Description of several new species of North American fishes. Journal of the Academy of Natural Science, Philadelphia, Pennsylvania, USA. 1(part 2):222-235, 359-368. (2)

306. Linton, E. 1898. Notes on cestode parasites of fishes. Proceedings of the United States National Museum 20:423-456. (15)

307. Linton, T. L. 1961. A study of fishes of the Arkansas and Cimarron rivers in the area of the proposed Keystone Reservoir. Oklahoma Fishery Research Laboratory Report Number 81. 30 pages. (1,13)

308. Litman, G. W., D. Frommel, and A. Rosenberg. 1971. Circular dichroic analysis of immunoglobulins in phylogenetic perspective. Biochimica et Biophysica Acta 236(3):647-654. (5)

309. Lockard, L. L., and R. R. Parsons. 1975. Some parasites of paddlefish (*Polyodon spathula*) from the Yellowstone River, Montana. Great Basin Naturalist 35(4):425-426. (15)

310. Ludwig, G. M., and R. Howmiller. 1971. A piscatorial pinocchio. Lore (Milwaukee Public Museum) 21(4):126-128. (7)

311. MacAlpin, A. 1947. *Paleopsephurus wilsoni,* a new polyodontid fish from the upper Cretaceous of Montana, with a discussion of allied fish, living and fossil. Contributions from the Museum of Paleontology, University of Michigan 6(8):167-234. (3)

312. Manges, D. E. 1959. Large impoundment investigations. Brush shelters. Tennessee Game and Fish Commission. Dingell-Johnson Project F-12-R-5, Job F. Final Report. 25 pages. (1)

313. Marchalonis, J. 1972. Conservatism in the evolution of immunoglobulin. Nature New Biology (London) 236(64):84-86. (5)

314. Marcoux, R. G. 1966. Occurrence of a melanistic paddlefish (*Polyodon spathula*) in Montana. Copeia 1966(4):876. (6)

315. Margenau, T. L. 1982. Evaluation of a power plant reservoir as a holding area for paddlefish brood stock. Master of Science Thesis. South Dakota State University, Brookings, South Dakota, USA. 30 pages. (6,10,14)

316. Martens, E. 1861. Ueber einen neuen *Polyodon* (*P. gladius*) aus dem Yantsekiang und Uber die sogenannten Glaspolypen. Monatsberichte der Deutschen Akademic der Wissenschaften zu Berlin. Pages 476-480. (16)

317. Mauduyt de la Varenne, P. J. E. 1774. Sur quelques objects du Regne animal apportes de la Louisiane. Rozier's Journal de Physique 4:382-397. (1,2,4)

318. McAllister, D. E., B. J. Parker, and P. M. McKee. 1981. Rare, endangered and extinct fishes in Canada. National Museum of Natural Sciences, Ottawa, Canada. 165 pages. (1,6)

319. McClurkin, I. T., A. K. Miksa, and L. A. Knight, Jr. 1981. The histology

and histochemistry of skeletal components in paddlefish, *Polyodon spathula*. Association of Southeastern Biologists Bulletin 28(2):47 (abstract). (4,5)

320. McCormick, L. M. 1892. Descriptive list of fishes of Lorain County, Ohio. Oberlin College Laboratory Bulletin Number 2. 34 pages. (1)

321. McLeran, V. 1973. Snaggin' for spoonbill. Kansas Fish and Game 30(1):3-6. (7,8)

322. McKinley, D. 1984. History of a neglected account of the paddlefish, *Polyodon spathula*. Copeia 1984 (1):201-204. (2)

323. Menhinick, E. F., T. M. Burton, and J. R. Bailey. 1974. An annotated checklist of the freshwater fishes of North Carolina. Journal of the Elisha Mitchell Scientific Society 90(1):24-50. (1)

324. Menon, M. D. 1950. The use of bones, other than otoliths, in determining the age and growth of fishes. Journal du Conseil, Conseil International pour l'Exploration de la Mer 16:311-335. (13)

325. Mensinger, G. C. 1971. Oklahoma commercial fisheries harvest summary, 1961-1969. Proceedings of the Oklahoma Academy of Science 1970 (51):23-28. (9)

326. Mensinger, G. C., and B. E. Brown. 1968. Live weight-dressed weight relationship for commercial fishes from four Oklahoma reservoirs. Proceedings of the Annual Conference Southeastern Association of Game and Fish Commissioners. 22:465-470. (9)

327. Merwin, J. 1973. Research gives paddlefish a chance. South Dakota Conservation Digest 40:12-14. (7,10,14)

328. Meyer, F. P. 1960. Life history of *Marsipometra hastata* and the biology of its host, *Polyodon spathula*. Doctoral Dissertation. Iowa State University, Ames, Iowa, USA. 145 pages. (6,9,11,12,13,15)

329. Meyer, F. P., and J. H. Stevenson. 1962. Studies on the artificial propagation of the paddlefish. The Progressive Fish-Culturist 24(2):65-67. (14)

330. Meyers, G. S. 1949. Salt tolerance of freshwater fish groups in relation to zoogeographical problems. Bijdragen Tot de Dierkunde 28:315-322. (5)

331. Michaletz, P. H. 1981. Paddlefish-zooplankton interactions in hatchery ponds. Master of Science Thesis. University of Missouri, Columbia, Missouri, USA. 94 pages. (12,13,14)

332. Michaletz, P. H., C. F. Rabeni, W. W. Taylor, and T. R. Russell. 1982. Feeding ecology and growth of young-of-the-year paddlefish in hatchery ponds. Transactions of the American Fisheries Society 111(6):700-709. (6,12)

333. Michaletz, P. H., C. F. Rabeni, W. W. Taylor, and T. R. Russell. 1983. Factors affecting *Daphnia* declines in paddlefish rearing ponds. The Progressive Fish-Culturist 45(2):76-80. (6,12)

334. Miksa, A. K. 1980. A histological and histochemical study of skeletal tissue of *Polyodon spathula*. Master of Science Thesis. University of Mississippi, Oxford, Mississippi, USA. 53 pages. (4)

335. Miles, R. L. 1971. Checklist of West Virginia fishes. West Virginia Department of Natural Resources, Charleston, West Virginia, USA. 5 pages. (1)

336. Miller, R. J., and H. W. Robinson. 1973. The fishes of Oklahoma. Oklahoma State University Press, Stillwater, Oklahoma, USA. 246 pages. (1,6)

337. Miller, R. R. 1972. Threatened freshwater fishes of the United States. Transactions of the American Fisheries Society 101(2): 239-252. (6)

338. Mims, S. D. 1984. Evaluation of *Daphnia* as a food for paddlefish, *Polyodon spathula,* fry under intensive culture conditions. Master of Science Thesis. Auburn University, Auburn, Alabama, USA. 60 pages. (6,9,11,12,13,14)

339. Mitchell, S. L., J. W. Clemens, and S. P. Hildreth. 1827. Notice of the spoonbill sturgeon, or paddlefish, of the Ohio, (*Polyodon feuille* of Lacepede). The American Journal of Science 12(2):201-205. (1,6)

340. Miyazaki, T., W. A. Rogers, and K. J. Semmens. 1983. Histopathological study on paddlefish: *Polyodon spathula* infested with juvenile *Hysterothylacium dollfusi* (Nematoda: Ascaroidea). Journal of Fish Diseases (in press). (15)

341. Morris, J., L. Morris, and L. Witt. 1972. The fishes of Nebraska. Nebraska Game and Parks Commission, Lincoln, Nebraska, USA. 98 pages. (1,6)

342. Moulton, J. M. 1957. A collection of drawings of fishes ascribed to J. P. Kirkland (1793-1877) in the Library of Bowdoin College. Breviora Number 80: 1-4 (Refers to a collection of drawings that contain original figures of Kirkland's accounts of *Polyodon folium* and *Platyrostra edentula.*) (2,4)

343. Munk, O. 1972. On the visual cells of some primitive fishes with particular regard to the classification of rods and cones. Videnskabelige Meddelelser fra Dansk Naturhistorisk Foreningi, Copenhagen. 132:25-30. (4)

344. Nachtrieb, H. F. 1902. The lateral line system of *Polyodon spathula.* Science (new series) 15(380):581-582. (4)

345. Nachtrieb, H. F. 1908. The primitive pores and the sensory ridges of the lateral line of *Polyodon spathula.* Science (new series) 27(702):914-915. (4)

346. Nachtrieb, H. F. 1910. The primitive pores of *Polyodon spathula* (Walbaum). Journal of Experimental Zoology 9(2):455-468. (4)

347. Nachtrieb, H. F. 1912. The lateral line system of *Polyodon.* Proceedings of the 7th International Zoological Congress, Boston, Massachusetts, USA. (1907):215-220. (4)

348. Nash, C. W. 1908. Checklist of the fishes of Ontario. Pages 1-122 *in* Manual of vertebrates of Ontario. Ontario Department of Education, Toronto, Canada. 247 pages. (1)

349. Needham, R. G. 1963. Harvest and life history of the paddlefish. Missouri Department of Conservation. Dingell-Johnson Project F-1-R-12, Work Plan 7, Job Number 3. Progress Report. 9 pages. (6,8,10,11)

350. Needham, R. G. 1964. Harvest and life history of the paddlefish. Missouri Department of Conservation. Dingell-Johnson Project F-1-R-13, Work Plan 7, Job Number 3. Progress Report. 18 pages. (6,8,10,11,14)

351. NEEDHAM, R. G. 1965A. Unique fishery. Missouri Conservationist 26(3):1. (7,8)

352. NEEDHAM, R. G. 1965B. Spawning of paddlefish induced by means of pituitary material. The Progressive Fish-Culturist 27(1):13-19. (14)

353. NEEDHAM, R. G. 1968. Paddlefish investigations. Montana Department of Fish and Game. Dingell-Johnson Project F-11-R-14, Job Number 2. Completion Report. 3 pages. (8,10)

354. NEEDHAM, R. G. 1969A. Paddlefish investigations. Montana Department of Fish and Game. Dingell-Johnson Project F-11-R-15, Job Number 2. Completion Report. 4 pages. (8,10)

355. NEEDHAM, R. G. 1969B. Paddlefish investigations. Montana Department of Fish and Game. Dingell-Johnson Project F-11-R-16, Job Number 2. Completion Report. 4 pages. (8,10)

356. NEEDHAM, R. G. 1970. Paddlefish investigations. Montana Department of Fish and Game. Dingell-Johnson Project F-11-R-17, Job Number 2-A. Completion Report. 4 pages. (8,10)

357. NEEDHAM, R. G. 1971. Paddlefish investigations. Montana Department of Fish and Game. Dingell-Johnson Project F-11-R-18, Job Number 2-A. Progress Report. 3 pages (8,10)

358. NEEDHAM, R. G. 1973A. Paddlefish investigations. Montana Department of Fish and Game. Dingell-Johnson Project F-11-R-19, Job Number 2-A. Progress Report. 3 pages. (8,10)

359. NEEDHAM, R. G. 1973B. Paddlefish investigations. Montana Department of Fish and Game. Dingell-Johnson Project F-11-R-20, Job Number 2-A. Progress Report. 5 pages. (8,10)

360. NEEDHAM, R. G. 1974A. Paddlefish investigations. Montana Department of Fish and Game. Dingell-Johnson Project F-11-R-22, Job Number 2-A. Progress Report. 6 pages. (8,10)

361. NEEDHAM, R. G. 1974B. Paddlefish investigations. Montana Department of Fish and Game. Dingell-Johnson Project F-11-R-21, Job Number 2-A. Progress Report. 1 page. (6,8,10)

362. NEEDHAM, R. G. 1976. Paddlefish investigations. Montana Department of Fish and Game. Dingell-Johnson Project F-11-R-23, Job Number 2-A. Progress Report. 4 pages. (8,10)

363. NEEDHAM, R. G. 1977A. Paddlefish investigations. Montana Department of Fish and Game. Dingell-Johnson Project F-11-R-24, Job Number 2-A. Progress Report. 4 pages. (8,10)

364. NEEDHAM, R. G. 1977B. Paddlefish investigations. Montana Department of Fish and Game. Dingell-Johnson Project F-11-R-25, Job Number 2-A. Progress Report. 4 pages. (6,8,10)

365. NEEDHAM, R. G. 1979. Paddlefish investigations. Montana Department of Fish, Wildlife, and Parks. Dingell-Johnson Project F-11-R-26, Job Number 2-A. Progress Report. 12 pages. (6,8,10)

366. NEEDHAM, R. G. 1980A. Paddlefish investigations. Montana Department of Fish, Wildlife, and Parks. Dingell-Johnson Project F-11-R-27, Job Number 2-A. Progress Report. 7 pages. (6,8,10)

367. NEEDHAM, R. G. 1980B. Paddlefish investigations. Montana Department of Fish, Wildlife, and Parks. Dingell-Johnson Project F-11-R-28, Job Number 2-A. Progress Report. 7 pages. (6,8,10)

368. NEEDHAM, R. G. 1981. Paddlefish investigations. Montana Department of Fish, Wildlife, and Parks. Dingell-Johnson Project F-11-R-29, Job Number 2-A. Progress Report. 7 pages. (6,8,10)

369. NEEDHAM, R. G. 1982. Paddlefish investigations. Montana Department of Fish, Wildlife, and Parks. Dingell-Johnson Project F-11-R-30, Job Number 2-A. Progress Report. 7 pages. (6,8,10)

370. NEEDHAM, R. G. 1983. Paddlefish investigations. Montana Department of Fish, Wildlife, and Parks. Dingell-Johnson Project F-11-R-31, Job Number 2-A. Progress Report. 5 pages. (6,8,10)

371. NEEDHAM, R. G. 1984. Paddlefish investigations. Montana Department of Fish, Wildlife, and Parks. Dingell-Johnson Project F-11-R-32, Job Number 2-A. Progress Report. 6 pages. (6,8,10)

372. NELSON, E. W. 1876. A partial catalogue of the fishes of Illinois. Illinois Museum of Natural History Bulletin 1(1):33-52. (1)

373. NELSON, J. S. 1976. Fishes of the world. John Wiley and Sons, New York, New York, USA. 416 pages. (1,16)

374. NELSON, W. R. 1961A. Report of fisheries investigations during the eighth year of impoundment of Fort Randall Reservoir, South Dakota, 1960. South Dakota Department of Game, Fish, and Parks. Dingell-Johnson Project F-1-R-10, Job Numbers 3-8. Progress Report. 29 pages. (6,8,10,14)

375. NELSON, W. R. 1961B. Report of fisheries investigations during the sixth year of impoundment of Gavins Point Reservoir, South Dakota, 1960. South Dakota Department of Game, Fish, and Parks. Dingell-Johnson Project F-1-R-10, Job Numbers 2-4. Progress Report. 56 pages. (8,10)

376. NELSON, W. R. 1962. Report of fisheries investigations during the ninth year of impoundment of Fort Randall Reservoir, South Dakota, 1961. South Dakota Department of Game, Fish, and Parks. Dingell-Johnson Project F-1-R-11, Job Numbers 3-6 and 9. Progress Report. 13 pages. (8,10)

377. NEW, J. G., AND D. BODZNICK. 1985. Segregation of electroreceptive and mesoreceptive lateral line afferents in the hind brain of Chondrostean fishes. Brain Research 336 (1985):89-98. (5)

378. NICHOLS, H. M. 1962. Reelfoot Lake investigations. Tennessee Game and Fish Commission. Dingell-Johnson Project F-5-R-10. Final Report. 57 pages. (1)

379. NICHOLS, J. T. 1916. A large *polyodon* from Iowa. Copeia 1916(34):65. (1,6)

380. NICHOLS, J. T. 1928. Chinese freshwater fishes in the American Museum of Natural History's collection. Bulletin of the American Museum of Natural History 58:1-62. (16)

381. NICHOLS, J. T. 1943. The fresh-water fishes of China. Natural History of Central Asia Volume 9. American Museum of Natural History. 322 pages. (16)

382. NIKOLSKII, G. V. 1961. Special ichthyology. Israel Program for Scientific Translations, Jerusalem. (National Technical Information Service,

Department of Commerce, Springfield, Virginia, USA.) 538 pages. (6)

383. Norman, J. R. 1975. A history of fishes. 3rd edition revised by P. H. Greenwood. John Wiley and Sons, New York, New York, USA. 467 pages. (6)

384. Norris, H. W. 1923. On the function of the paddle of the paddlefish. Proceedings of the Iowa Academy of Science 30:135-137. (4,6)

385. Norris, H. W. 1925. Observations upon the peripheral distribution of the cranial nerves of certain ganoid fishes (*Amia, Lepisosteus, Polyodon, Scaphyrhynchus* and *Acipsenser*). Journal of Comparative Neurology 39(3):333-345. (4)

386. O'Donnell, D. J. 1935. Annotated list of the fishes of Illinois. Illinois Natural History Survey Bulletin 20(5):473-500. (1)

387. Ohio Department of Natural Resources. No date. Paddlefish in Ohio. Division of Wildlife Publication 186. Columbus, Ohio, USA. (1,7)

388. Oklahoma City Times. 1951. Record weight spoonbill catfish. Oklahoma Game Fish News 7(2):14. (1,7)

389. Olson, S. 1971. Shadows in the river. NEBRASKAland 49(1):34-37. (7)

390. Pasch, R. W., and C. M. Alexander. 1986. Effects of commercial fishing on paddlefish populations. Pages 46-53 *in* J. G. Dillard, L. K. Graham, and T. R. Russell, editors. The paddlefish: status, management and propagation. North Central Division, American Fisheries Society, Special Publication Number 7. 159 pages. (6,9,10)

391. Pasch, R. W., P. A. Hackney, and C. M. Alexander. 1983. Estimated abundance of young-of-the-year paddlefish in Old Hickory Reservoir, Tennessee. Tennessee Valley Authority, Office of Natural Resources, Division of Air and Water Resources, Knoxville, Tennessee, USA. Progress Report. 26 pages. (6,10)

392. Pasch, R. W., P. A. Hackney, and J. A. Holbrook. 1978. Ecology of the paddlefish, *Polyodon spathula* (Walbaum), in Old Hickory Reservoir, Tennessee, with emphasis on first year life history and the effects of impingement at Gallatin Steam Plant on the population. Tennessee Valley Authority, Division of Water Resources, Knoxville, Tennessee, USA. Developmental report. 54 pages. (6,8,10,13)

393. Pasch, R. W., P. A. Hackney, and J. A. Holbrook. 1980. Ecology of paddlefish in Old Hickory Reservoir, Tennessee, with emphasis on first-year life history. Transactions of the American Fisheries Society 109(2):157-167. (6,9,11,13)

394. Patterson, C. 1975. The distribution of Mesozoic freshwater fishes. Memoires du Museum National d'Historic Naturelle (France) Serie A Zoologie 88:156-174. (3)

395. Pehrson, T. 1944. Some observations on the development and morphology of the dermal bones in the skull of *Acipenser* and *Polyodon*. Acta Zoologica (Stockholm) 24:27-48. (4)

396. Peterson, D. C., and C. M. Alexander. 1984. An evaluation of the Cherokee Reservoir paddlefish population. Tennessee Valley Authority, Office of

Natural Resources, Division of Air and Water Resources, Knoxville, Tennessee, USA. Progress Report. 7 pages. (6,9,10)

397. PFLIEGER, W. L. 1971. A distributional study of Missouri fishes. University of Kansas Museum of Natural History Publication 20(3):225-570. (1,6)

398. PFLIEGER, W. L. 1975. The fishes of Missouri. Missouri Department of Conservation, Jefferson City, Missouri, USA. 343 pages. (1,6)

399. PHILLIPS, G. L., W. D. SCHMID, AND J. C. UNDERHILL. 1982. Fishes of the Minnesota Region. University of Minnesota Press, Minneapolis, Minnesota, USA. 248 pages. (1,6)

400. PING, C. 1931. Preliminary notes on the fauna of Nanking. Contributions from the Biological Laboratory of the Science Society of China, Zoological Series 7(4):173-201. (16)

401. POLK, B. H. 1879. Paddle-fish or shovel-nose sturgeon. Forest and Stream 11(24):482. (6,7)

402. POLLARA, B., J. FINSTAD, AND R. A. GOOD. 1966. The phylogenetic development of immunoglobulins. Pages 88-97 *in* R. T. Smith, P. A. Miescher, and R. A. Good, editors. Phylogeny of immunity. University of Florida Press, Gainesville, Florida, USA. 276 pages. (5)

403. PURKETT, C. A., JR. 1959. An investigation of a fishery for paddlefish with attention to the life history of this fish. Missouri Department of Conservation. Dingell-Johnson Project F-1-R-8, Work Plan 8, Job Number 1. Progress Report. 5 pages. (6,8,10,11,13)

404. PURKETT, C. A., JR. 1960. An investigation of a fishery for paddlefish with attention to the life history of this fish. Missouri Department of Conservation. Dingell-Johnson Project F-1-R-9, Work Plan 8, Job Number 1. Progress Report. 8 pages. (6,8,10,11,13)

405. PURKETT, C. A., JR. 1961A. Fire in the hole. Missouri Conservationist 22(4):1. (7,8)

406. PURKETT, C. A., JR. 1961B. A day to remember. Missouri Conservationist 22(5):10. (7,11)

407. PURKETT, C. A., JR. 1961C. Reproduction and early development of the paddlefish. Transactions of the American Fisheries Society 90(2):125-129. (6,11,14)

408. PURKETT, C. A., JR. 1961D. An investigation of a fishery for paddlefish with attention to the life history of this fish. Missouri Department of Conservation. Dingell-Johnson Report F-1-R-10, Work Plan 8, Job Number 1. Progress Report. 14 pages. (6,8,10,11,13)

409. PURKETT, C. A., JR. 1963A. The paddlefish fishery of the Osage River and the Lake of the Ozarks, Missouri. Transactions of the American Fisheries Society 92(3):239-244. (6,8,10,13)

410. PURKETT, C. A., JR. 1963B. Artificial propagation of the paddlefish. The Progressive Fish-Culturist 25(1):31-33. (14)

411. RADFORTH, I. 1944. Some considerations on the distribution of fishes in Ontario. Contributions from the Royal Ontario Museum of Zoology Number 25. 116 pages. (1)

412. Radisson, P. E. 1961. The explorations of Pierre Espirit Radisson, edited by A. T. Adams. Ross and Haines, Minneapolis, Minnesota, USA. 258 pages. (1,2,4)

413. Raikova, E. V., V. C. Suppes, and G. L. Hoffman. 1979. The parasitic coelenterate, *Polypodium hydriforme ussov,* from the eggs of the American acipenseriform *Polyodon spathula.* Journal of Parasitology 65(5):804-810. (15)

414. Rafinesque, C. S. 1820. Ichthyologia Ohiensis, or Natural History of the fishes inhabiting the River Ohio and its tributary streams, preceded by a physical description of the Ohio and its branches. Lexington, Kentucky, USA. 90 pages. (facsimile report, Arno, New York, USA; 1970). (2)

415. Regier, G. 1974. Vanishing freshwater giants. National Wildlife 12(5):25-29. (7)

416. Regier, G. 1977a. Native fish in troubled waters. Audubon 79(1):18-41. (6,7)

417. Regier, G. 1977b. The incredible paddlefish. Field and Stream 82(7):62,82,84. (6,7,8)

418. Rehwinkel, B. J. 1975. The fishery for paddlefish at Intake, Montana, during 1973 and 1974. Master of Science Thesis. Montana State University, Bozeman, Montana, USA. 36 pages. (6,8,10,13)

419. Rehwinkel, B. J. 1978. The fishery for paddlefish at Intake, Montana during 1973 and 1974. Transactions of the American Fisheries Society 107(2):263-268. (6,8,10,13)

420. Richards, K. R. 1982. Study of the fishery of the Upper Osage Arm of Lake of the Ozarks, Missouri. Contract report DACW41-81-C-0103. Final Report. United States Army Corps of Engineers, Kansas City, Missouri, USA. 19 pages. (6,8,10)

421. Richards, K. R., R. J. Dent, and W. H. Dieffenbach. 1986. Fisheries problems associated with the Truman Dam pumped storage hydroelectric project in West Central Missouri. Pages 247-254 *in* G. E. Hall and M. J. Van Den Avyle, editors. Reservoir fisheries management: strategies for the 80's. Reservoir Committee, Southern Division American Fisheries Society. Bethesda, Maryland, USA. (1,6,8,10)

422. Riggs, C. D., and G. A. Moore. 1950. Some new records of paddlefish and sturgeon from Oklahoma. Proceedings of the Oklahoma Academy of Science 30(1949):16-18. (1)

423. Robinson, J. W. 1966. Observations on the life history, movement, and harvest of the paddlefish, *Polyodon spathula*, in Montana. Proceedings of the Montana Academy of Science 26:33-44. (6,8,10,11,13)

424. Robinson, R. 1979. Fishing for paddlefish. The Drum and Croaker 19(2):23-24. (6,7)

425. Rosen, R. A. 1976. Distribution, age and growth, and feeding ecology of paddlefish (*Polyodon spathula*) in unaltered Missouri River, South Dakota. Master of Science Thesis. South Dakota State University, Brookings, South Dakota, USA. 95 pages. (1,6,10,11,12,13)

426. Rosen, R. A., and D. C. Hales. 1980. Occurrence of scarred paddlefish in the

Missouri River, South Dakota-Nebraska. The Progressive Fish-Culturist 40(2):82-85 (4,6,8)

427. ROSEN, R. A., AND D. C. HALES. 1981. Feeding of paddlefish, *Polyodon spathula*. Copeia 1981 (2):441-455. (4,6,12)

428. ROSEN, R. A., AND D. C. HALES. 1982. Occurrence of a blind paddlefish (*Polyodon spathula*). Copeia 1982 (1):212-214. (4,6,8)

429. ROSEN, R. A., D. C. HALES, AND D. G. UNKENHOLZ. 1982. Biology and exploitation of paddlefish in the Missouri River below Gavins Point Dam. Transactions of the American Fisheries Society 111(2):216-222. (1,6,8,10,13)

430. ROSTLUND, E. 1951. Three early historical reports of North American freshwater fishes. Copeia 1951 (4):295-296. (1,2,4)

431. ROSTLUND, E. 1952. Freshwater fish and fishing in native North America. University of California Publications in Geological Sciences 9:1-314. (1)

432. RUELLE, R., AND P. L. HUDSON. 1977. Paddlefish (*Polyodon spathula*): growth and food of young of the year and a suggested technique for measuring length. Transactions of the American Fisheries Society 106(6):609-613. (10,12)

433. RUHR, C. E. 1956. Nature page (paddlefish). Tennessee Conservationist 22(4):23-24. (7)

434. RUPP, L. 1974. Paddlefish . . . What about tomorrow? Nebraska Game & Parks Commission, Lincoln, Nebraska, USA. (Pamphlet on sport snagging regulation changes in Nebraska) (8,10)

435. RUPP, L. 1982. Notes on Nebraska fauna — paddlefish. NEBRASKAland 60(3):50. (6,7)

436. RUSSELL, T. R. 1965. Harvest and life history of the paddlefish. Missouri Department of Conservation. Dingell-Johnson Project F-1-R-14, Work Plan 7, Job Number 3. Progress Report. 6 pages. (6,8,10,14)

437. RUSSELL, T. R. 1966. Harvest and life history of the paddlefish. Missouri Department of Conservation. Dingell-Johnson Project F-1-R-15, Work Plan 7, Job Number 3. Progress Report. 7 pages. (6,8,10,14)

438. RUSSELL, T. R. 1967. Life history and artificial propagation of the paddlefish. Missouri Department of Conservation. Dingell-Johnson Project F-1-R-16, Work Plan 7, Job Number 3. Progress Report. 9 pages. (6,8,10,14)

439. RUSSELL, T. R. 1968. Life history and artificial propagation of the paddlefish. Missouri Department of Conservation. Dingell-Johnson Project F-1-R-17, Work Plan 7, Job Number 3. Progress Report. 9 pages. (6,8,10,14)

440. RUSSELL, T. R. 1969. Life history and artificial propagation of the paddlefish. Missouri Department of Conservation. Dingell-Johnson Project F-1-R-18, Work Plan 7, Job Number 3. Progress Report. 11 pages. (6,8,10,14)

441. RUSSELL, T. R. 1970. Life history and artificial propagation of the paddlefish. Missouri Department of Conservation. Dingell-Johnson Project F-1-R-19, Work Plan 7, Job Number 3. Progress Report. 12 pages. (6,8,10,14)

442. RUSSELL, T. R. 1971. A study of artificial propagation of the paddlefish. Missouri Department of Conservation. Dingell-Johnson Project F-1-R-20, Study S-5, Job Number 1. Progress Report. 11 pages. (14)

443. Russell, T. R. 1972. Age and growth of the paddlefish. Missouri Department of Conservation. Dingell-Johnson Project F-1-R-21, Study S-4, Job Number 1. Final Report. 9 pages. (6,8,13)

444. Russell, T. R. 1973. A study of artificial propagation of the paddlefish. Missouri Department of Conservation. Dingell-Johnson Project F-1-R-21, Study S-5, Job Numbers 1 and 2. Final Report. 16 pages. (14)

445. Russell, T. R. 1982. Paddlefish. Missouri Conservationist 43(3):11-13. (6,7)

446. Russell, T. R. 1986. Biology and life history of the paddlefish — a review. Pages 2-20 *in* J. G. Dillard, L. K. Graham and T. R. Russell, editors. The paddlefish: status, management and propagation. North Central Division, American Fisheries Society, Special Publication Number 7. 159 pages. (6,10,11,12,13)

447. Russell, T. R., L. K. Graham, D. M. Carlson, and E. J. Hamilton. 1975. Maintenance of the Osage River paddlefish fishery. Missouri Department of Conservation, Columbia, Missouri, USA. 13 pages. (1,6,8,10,11,12,13,14)

448. Russell, T. R., L. K. Graham, D. M. Carlson, and E. J. Hamilton. 1976. Maintenance of the Osage River paddlefish fishery. Missouri Department of Conservation, Columbia, Missouri, USA. Progress Report. 27 pages. (1,6,8,10,11,12,13,14)

449. Russell, T. R., L. K. Graham, D. M. Carlson, and E. J. Hamilton. 1977. Maintenance of the Osage River paddlefish fishery. Missouri Department of Conservation, Columbia, Missouri, USA. Progress Report. 29 pages. (1,6,8,10,11,12,13,14)

450. Russell, T. R., L. K. Graham, D. M. Carlson, and E. J. Hamilton. 1980. Maintenance of the Osage River - Lake of the Ozarks paddlefish fishery. Missouri Department of Conservation, Columbia, Missouri, USA. Final Report. 33 pages. (1,6,8,10,11,12,13,14)

451. Sample, D. W. 1980. Estimated commercial fish and mussel harvest from the Tennessee Valley. Tennessee Valley Authority, Office of Natural Resources and Economic Development, Knoxville, Tennessee, USA. 15 pages. (9)

452. Schmidt, G. D., P. D. Leiby, and D. C. Kritsky. 1974. Studies on helminths of North Dakota: Part 8. Nematodes from the paddlefish, *Polyodon spathula* (Walbaum), including *Thynnascaris dollfusi,* new species, (Nematoda: Stomachidae). Canadian Journal of Zoology 52(2):261-263. (15)

453. Schulmach, J. C., G. Gould, and C. L. Groen. 1975. Relative abundance and distribution of fishes in the Missouri River, Gavins Point Dam to Rulo, Nebraska. Proceedings of the South Dakota Academy of Science 54:194-222. (6,8)

454. Scott, W. B. 1967. Freshwater fishes of Eastern Canada, 2nd edition. University of Toronto Press, Toronto, Canada. 137 pages. (1)

455. Semmens, K. J. 1982. Production of fingerling paddlefish (*Polyodon spathula*) in earthen ponds. Master of Science Thesis. Auburn University, Auburn, Alabama, USA. 54 pages. (14)

456. Semmens, K. J., and W. L. Shelton. 1986. Opportunities in paddlefish

aquaculture. Pages 106-113 *in* J. G. Dillard, L. K. Graham and T. R. Russell, editors. The paddlefish: Status, management and propagation. North Central Division, American Fisheries Society, Special Publication Number 7. 159 pages. (10,14)

457. SETTE, O. E. 1925. Fisheries industries of the United States, 1923. Appendix 4, Report of the United States Commissioners of Fisheries. 1924:141-359. (9)

458. SEWERTZOFF, A. N. 1923. The place of the cartilaginous ganoids in the system and the evolution of the Osteichthyes. Journal of Morphology 38(1):105-145. (2,3)

459. SHAW, G. 1803-1804. Piscus. *in* General zoology. Four parts in volumes 4-5. Kearsley, London. (1,2,4)

460. SHELTON, W. L., AND F. MERIWETHER. 1978. Adult paddlefish growth and maturation. Fisheries Research Annual Report. Auburn University, Auburn, Alabama, USA. 5:144-148. (6,10,13)

461. SHELTON, W. L., AND R. O. SMITHERMAN. 1984. Exotic fishes in warmwater aquaculture. Pages 262-301 *in* W. R. Courtenay, Jr. and J. R. Stauffer, Jr., editors. Distribution, biology and management of exotic fishes. The Johns Hopkins University Press, Baltimore, Maryland, USA. 430 pages. (6,14)

462. SHIELDS, J. T. 1955. Report of fisheries investigations during the second year of impoundment at Fort Randall Reservoir, South Dakota, 1954. South Dakota Department of Game, Fish, and Parks. Dingell-Johnson Project F-1-R-4. Progress Report. 100 pages. (8,10)

463. SHIELDS, J. T. 1956. Report of fisheries investigations during the third year of impoundment of Fort Randall Reservoir, South Dakota, 1955. South Dakota Department of Game, Fish, and Parks. Dingell-Johnson Project F-1-R-5. Progress Report. 91 pages. (8,10)

464. SHIELDS, J. T. 1957. Report of fisheries investigations during the fourth year of impoundment of Fort Randall Reservoir, South Dakota, 1956. South Dakota Department of Game, Fish, and Parks. Dingell-Johnson Project F-1-R-6. Progress Report. 54 pages. (8,10)

465. SHIELDS, J. T. 1958. Report of the fisheries investigations during the fifth year of impoundment of Fort Randall Reservoir, South Dakota, 1957. South Dakota Department of Game, Fish, and Parks. Dingell-Johnson Project F-1-R-7. Progress Report. 29 pages. (8,10)

466. SIMER, P. H. 1929. Fish trematodes from the lower Tallahatchie River. American Midland Naturalist 11(12):563-588. (15)

467. SIMER, P. H. 1930. A preliminary study of the cestodes of the spoonbill, *Polyodon spathula* (Walbaum). Transactions of the Illinois State Academy of Science 22:139-145. (15)

468. SLASTENENKO, E. P. 1958. The freshwater fishes of Canada. Kiev Printers, Toronto, Canada. 385 pages. (1,6)

469. SMITH, H. M. 1898. Statistics of the fisheries of the interior waters of the United States. Report of the United States Commission of Fish and Fisheries for the year ending June 30, 1896. Part 12. Pages 489-574. (9)

470. SMITH, P. W. 1965. A preliminary annotated list of the lampreys and fishes of Illinois. Illinois Natural History Survey Biological Notes 54. 12 pages. (1)

471. SMITH, P. W. 1979. The fishes of Illinois. University of Illinois Press, Urbana, Illinois, USA. 314 pages. (1,6)

472. SMITH, P. W., A. C. LOPINOT, AND W. L. PFLIEGER. 1971. A distributional atlas of Upper Mississippi River fishes. Illinois Natural History Survey Biological Notes Number 73. 19 pages. (1)

473. SMITH, R. J. 1958. South Dakota's peculiar paddlefish. Sports Afield 139:48-49, 107-109. (7,8)

474. SMITH-POLLARD, L. 1980. Paddlefish. Outdoor Oklahoma. 36(3):16-20. (6,7)

475. SMITH-VANIZ, W. F. 1968. Freshwater fishes of Alabama. Alabama Agricultural Experiment Station, Auburn University, Auburn, Alabama, USA. 211 pages. (1,6)

476. SNYDER, D. E. 1980. Characteristics for distinguishing the protolarvae of the paddlefish and sturgeon. Early Life History Section Newsletter, American Fisheries Society, Bethesda, Maryland, USA. 1(2):9-10. (4)

477. SOUTHALL, P. D. 1982. Paddlefish movement and habitat use in the upper Mississippi River. Master of Science Thesis. Iowa State University, Ames, Iowa, USA. 100 pages. (1,11)

478. SOUTHALL, P. D., AND W. A. HUBERT. 1984. Habitat use by adult paddlefish in the upper Mississippi River. Transactions of the American Fisheries Society 113(2):125-131. (1,6)

479. SPARROWE, R. D. 1977. Harry S. Truman Dam and Reservoir — a case history of problems in fish and wildlife coordination. Transactions of the North American Wildlife Natural Resource Conference. 42:42-55. (8,10)

480. SPARROWE, R. D. 1986. Threats to paddlefish habitat. Pages 36-45 *in* J. G. Dillard, L. K. Graham, and T. R. Russell, editors. The paddlefish: status, management and propagation. North Central Division, American Fisheries Society, Special Publication Number 7. 159 pages. (10)

481. SPRAGUE, J. W. 1959A. Report of fisheries investigations during the sixth year of impoundment of Fort Randall Reservoir, South Dakota, 1958. South Dakota Department of Game, Fish, and Parks. Dingell-Johnson Project F-1-R-8, Job Numbers 1-5. Progress Report. 29 pages. (8,10)

482. SPRAGUE, J. W. 1959B. Report of fisheries investigations during the fourth year of impoundment of Gavins Point Reservoir, South Dakota, 1958. South Dakota Department of Game, Fish, and Parks. Dingell-Johnson Project F-1-R-8, Job Numbers 3 and 7. Progress Report. 47 pages. (8,10)

483. SPRAGUE, J. W. 1960. Report of fisheries investigations during the fifth year of impoundment of Gavins Point Reservoir, South Dakota, 1959. South Dakota Department of Game, Fish, and Parks. Dingell-Johnson Project F-1-R-9, Job Numbers 3 and 7. Progress Report. 47 pages. (8,10)

484. SPRAGUE, J. W. 1961. Report of fisheries investigations during the seventh year of impoundment of Fort Randall Reservoir, South Dakota, 1959. South Dakota Department of Game, Fish, and Parks. Dingell-Johnson

Project F-1-R-9, Job Numbers 5-8. Progress Report. 47 pages. (8,10)

485. STALIONS, B. 1980. The spoonbill. Skin Diver 29(2):56,58. (7)

486. STARRETT, W. C., AND W. A. FRITZ. 1965. A biological investigation of the fishes of Lake Chautauqua, Illinois. Illinois Natural History Survey Bulletin 29(1):19. (1)

487. STEWART, G. A., AND P. JOHNS. 1976. Empirical and theoretical relationships between the sedimentation coefficient and molecular weights of various proteins, with particular reference to the immunoglobulins. Journal of Immunological Methods 10(2-3):219-229. (5)

488. STEWART, P. A. 1984. Paddlefish investigations. Montana Department of Fish, Wildlife, and Parks. Dingell-Johnson Project F-30-R-20, Job Number 2-A. Progress Report. 11 pages. (6,8,10)

489. STOCKARD, C. R. 1907. Observations on the natural history of *Polyodon spathula*. American Naturalist 41(492):753-766. (6,9,11,12)

490. STONE, N. M. 1981. Growth of male and female *Tilapia nilotica* in ponds and cages. Master of Science Thesis. Auburn University, Auburn, Alabama, USA. 19 pages. (reports one season's growth of pond reared paddlefish) (13,14)

491. STORY, H. 1964. The olfactory bulbar formation and related nuclei of the paddlefish (*Polyodon spathula*). Journal of Comparative Neurology 123:285-297. (4)

492. SUPPES, C. V., AND F. P. MEYER. 1975. Polypodium species (Coelenterata) infection of paddlefish (*Polyodon spathula*) eggs. Journal of Parasitology 61(4):772-774. (15)

493. SWINGLE, H. A. 1965. Growth rates of paddlefish receiving supplemental feeding in fertilized ponds. The Progressive Fish-Culturist 27(4):220. (14)

494. SWINGLE, H. S. 1965. Management techniques for public fishing waters. Management of new species of fish for Alabama. Alabama Department of Conservation and Natural Resources, Division of Game and Fish. Dingell-Johnson Project F-10-R-6, Job Number 1. Progress Report. 16 pages. (1,10)

495. SWINGLE, W. E. 1965. Length-weight relationships of Alabama fishes. Auburn University Agricultural Experiment Station Zoology-Entomology Series. Fisheries Department, Auburn, Alabama, USA. 3:1-87. (13)

496. TATARKO, K. 1937. The maxillary and the hyoid arch and the apparatus of the operculum in the Polyodontidae. Trudy Instituta Zoolgii Bioloii (Kiev) 16:47-76. (4,16)

497. TARZWELL, C. M., AND P. BRYAN. 1945. Changes in the commercial fishery on the Tennessee River. Journal of the Tennessee Academy of Science 20(1):49-54. (9)

498. TCHANG, T. L. 1928. A review of the fishes of Nanking. Contributions from the Biological Laboratory of the Science Society of China 4(4):42 pages. (16)

499. TENNESSEE VALLEY AUTHORITY. 1945. Annual report, Biological Readjustment Division, fiscal year 1945. Tennessee Valley Authority, Office of

Natural Resources and Economic Development, Knoxville, Tennessee, USA. 43 pages. (9)

500. TENNESSEE VALLEY AUTHORITY. 1974. Annual Report. Division of Forestry, Fisheries, and Wildlife, Norris, Tennessee, USA. 40 pages. (9)

501. TENNESSEE VALLEY AUTHORITY. 1976. Effects of impingement at Gallatin Steam Plant on the populations of fish in Old Hickory Reservoir. Pages 1-7 *in* Effects of Gallatin Steam Plant cooling water intake on the fish populations of Old Hickory Reservoir. Tennessee Valley Authority, Office of Natural Resources, Division of Air and Water Resources, Knoxville, Tennessee, USA. (6,10)

502. THACHER, J. K. 1877. Ventral fins of ganoids. Transactions of the Connecticut Academy of Art and Sciences. 4:233-242. (4)

503. THIES, B. 1981. Finding new homes for Missouri's ancient fish. Fins and Feathers (Missouri) 3(3):66-68. (6,7)

504. THOMPSON, D. H. 1933. The finding of very young *Polyodon.* Copeia 1933 (1):31-33. (6)

505. THOMPSON, D. H. 1934. Relative growth in *Polyodon.* Illinois Natural History Survey Biological Notes. Number 2. 8 pages. (13)

506. THORSON, T. D. 1961. The partitioning of body water in Osteoichthyes — phylogenetic and ecological implications in aquatic invertebrates. Biological Bulletin (Woods Hole). 120(2):238-254. (5)

507. THWAITES, R. G., EDITOR. 1896-1901. The Jesuit relations and allied documents 1610-1791. Burrows Brothers, Cleveland, Ohio, USA. Volume 59. (1,2,4)

508. TOMLJANOVICH, D. A., J. A. HEUER, AND R. L. SCHWEINFORTH. 1977. A laboratory study on the swimming ability of juvenile paddlefish (*Polyodon spathula*). Tennessee Valley Authority, Office of Natural Resources and Water Resources, Knoxville, Tennessee, USA. Progress Report. 15 pages. (5,6)

509. TONER, G. C. 1933. Annotated list of fishes of Georgian Bay. Copeia 1933(3):133-140. (1)

510. TONDREAU, R. E. 1979. Ichthyoplankton distribution in the Missouri River. Master of Science Thesis. University of Iowa, Iowa City, Iowa, USA. 68 pages. (6,11)

511. TOWNSEND, C. H. 1902. Statistics of the fisheries of the Mississippi River and tributaries. Appendix 15, Report of the United States Commission of Fish and Fisheries. 27(1901):659-740. (9)

512. TRAUTMAN, M. B. 1981. The fishes of Ohio. Ohio State University Press, Columbus, Ohio, USA. 782 pages. (1,6)

513. TRAVELL, W. 1914. Paddlefish found in mud. Copeia 1914 (4):4. (1)

514. TURNER, W. R. 1959. Pre-impoundment surveys of Kentucky streams. Kentucky Department of Fish and Wildlife Resources, Fisheries Bulletin. Pages 24-43. (1)

515. TUTEN, J. S. 1977. Polyculture of red swamp crayfish (*Procambarus clarkii*),

channel catfish (*Ictalurus punctatus*), bigmouth buffalo (*Ictiobus cyprinellus*), and paddlefish (*Polyodon spathula*). Master of Science Thesis. Louisiana State University, Baton Rouge, Louisiana, USA. 47 pages. (14)

516. Tuten, J. S., and J. W. Avault, Jr. 1981. Growing red swamp crayfish (*Procambarus clarkii*) and several North American fish species together. The Progressive Fish-Culturist 43(2):97-99. (14)

517. Unkenholz, D. G. 1976. Investigation of paddlefish populations in South Dakota and development of management plans, 1976. South Dakota Department of Game, Fish, and Parks. Dingell-Johnson Project F-15-R-12, Study Number 9, Job Numbers 3, 5, and 7. Progress Report. 19 pages. (6,8,10,12,13,14)

518. Unkenholz, D. G. 1977. Investigation of paddlefish populations in South Dakota and development of management plans, 1977. South Dakota Department of Game, Fish, and Parks. Dingell-Johnson Project F-15-R-13, Study Number 9, Job Numbers 3, 5, 7. Progress Report. 34 pages. (6,8,10,12,13,14)

519. Unkenholz, D. G. 1979a. Investigation of the paddlefish population and recruitment in the Missouri River below Gavins Point Dam. South Dakota Department of Game, Fish, and Parks. Dingell-Johnson Project F-15-R-14, Study Number 9, Job Number 5. Progress Report Number 79-2. 4 pages. (6,8,10)

520. Unkenholz, D. G. 1979b. Investigation of paddlefish populations in South Dakota and development of management plans. South Dakota Department of Game, Fish, and Parks. Dingell-Johnson Project F-15-R-14, Study Number 9, Job Numbers 3 and 7. Progress Report Number 79-1. 25 pages. (6,8,10,12,14)

521. Unkenholz, D. G. 1980a. Paddlefish spawning movements and reproductive success in the Missouri River below Fort Randall Dam, 1979. South Dakota Department of Game, Fish, and Parks. Progress Report Number 80-3. 13 pages. (6,8,10,13)

522. Unkenholz, D. G. 1980b. Investigation of paddlefish populations and recruitment in the Missouri River below Gavins Point Dam. South Dakota Department of Game, Fish, and Parks. Dingell-Johnson Project F-15-R-15, Study Number 9, Job Number 5, Progress Report Number 80-2. 7 pages. (6,8,10,13)

523. Unkenholz, D. G. 1981a. Investigation of the paddlefish population and recruitment in the Missouri River below Gavins Point Dam. South Dakota Department of Game, Fish, and Parks. Dingell-Johnson Project F-15-R-16, Study Number 9, Job Number 5. Progress Report Number 81-2. 5 pages. (6,8,10)

524. Unkenholz, D. G. 1981b. Paddlefish spawning movements and reproductive success in the Missouri River below Fort Randall Dam, 1980. South Dakota Department of Game, Fish, and Parks. Progress Report Number 81-3. 17 pages. (6,8,10,13)

525. Unkenholz, D. G. 1981c. Big Movers. South Dakota Conservation Digest 48(2):2-3. (6,7)

526. Unkenholz, D. G. 1982a. Paddlefish spawning movements and reproductive success in the Missouri River below Fort Randall Dam, 1979-81. South Dakota Department of Game, Fish, and Parks. Progress Report Number 82-3. 17 pages. (6,8,10,11)

527. Unkenholz, D. G. 1982b. Investigation of the paddlefish populations and recruitment in the Missouri River below Gavins Point Dam. South Dakota Department of Game, Fish, and Parks. Dingell-Johnson Project F-15-R-17, Study Number 9, Job Number 5. Progress Report Number 82-9. 15 pages. (6,8,10,11,13)

528. Unkenholz, D. G. 1986. Effects of dams and other habitat alterations on paddlefish sport fisheries. Pages 54-61 *in* J. G. Dillard, L. K. Graham, and T. R. Russell, editors. The paddlefish: status, management and propagation. North Central Division, American Fisheries Society, Special Publication Number 7. 159 pages. (6,8,10)

529. Van der Horst, C. J. 1925. The myclencephalic gland of *Polyodon, Acipenser,* and *Amia*. Proceedings of the Academy of Science, Amsterdam 28:432-442. (4,5)

530. Van Eeckhout, G. 1978. Investigations of selected fish populations by the mark-recapture method. North Dakota State Game and Fish Department. Dingell-Johnson Project F-2-24 and 25, Study 4, Job Number 4-A. Progress Report. 7 pages. (8,9,10)

531. Van Eeckhout, G. 1980. Investigations of selected fish populations by the mark-recapture method. North Dakota State Game and Fish Department. Dingell-Johnson Project F-2-R-27, Study 4, Job Number 4-A. Final Report. 25 pages. (6,8,10)

532. Vance, J. M. 1971. Paddlefish out progressed. Missouri Conservationist 32(8):1-3. (6,7)

533. Vance, J. M. 1979. The gentle giant. Rod and Reel 6:40-43. (6,7)

534. Vasetskiy, S. G. 1971. Fishes of the family Polyodontidae. Journal of Ichthyology (USSR) 11(1):18-31. (2,4,5,6,8,9,10,11,12,13,14,15,16)

535. Wagner, G. 1904. Notes on *Polyodon*. Science (new series) 19 (483):554-555. (6)

536. Wagner, G. 1908. Notes on the fish fauna of Lake Pepin. Transactions of the Wisconsin Academy of Science 16 (part 1):23-37. (9)

537. Walbaum, J. J. 1792. Petri Artedi renovati, bibliotheca et philosophia ichthyologica. Ichthyologiae part 3. Grypeswaldiae. Page 522. *in* P. Artedi 1788-1793. 5 parts in 3 volumes. 723 pages. (1,2,4)

538. Walburg, C. H. 1964. Fish population studies, Lewis and Clark Lake, Missouri River, 1956 to 1962. United States Fish and Wildlife Service, Special Scientific Report-Fisheries 482. 27 pages. (6,11)

539. Walburg, C. H. 1971. Loss of young fish in reservoir discharge and year-class survival, Lewis and Clark Lake, Missouri River. Pages 441-448 *in* G. E. Hall, editor. Reservoir fisheries and limnology. American Fisheries

Society Special Publication Number 8, Bethesda, Maryland, USA. 511 pages. (11)

540. WALBURG, C. H., G. L. KAISER, AND P. L. HUDSON. 1971. Lewis and Clark Lake tailwater biota and some relations of the tailwater and reservoir fish populations. Pages 449-467 *in* G. E. Hall, editor. Reservoir fisheries and limnology. American Fisheries Society Special Publication Number 8, Bethesda, Maryland, USA. 511 pages. (11)

541. WALLUS, R. 1983. Paddlefish reproduction in the Cumberland and Tennessee river systems. Tennessee Valley Authority, Office of Natural Resources, Division of Air and Water Resources, Knoxville, Tennessee, USA. Progress Report. 37 pages. (6,10,11,13)

542. WEBB, D. H., AND M. E. SISK. 1975. The fishes of West Kentucky. III. The fishes of Bayou de Chien. Transactions of the Kentucky Academy of Science 36(3-4):63-70. (1)

543. WEED, A. C. 1925. Feeding the paddlefish. Copeia 1925, No. 146:67-68. (14)

544. WEINHEIMER, P. F., J. MESTECKY, AND R. T. ACTON. 1971. Species distribution of J Chain. Journal of Immunology 107(4): 1211-1212. (5)

545. WEISEL, G. F. 1973. Anatomy and histology of the digestive system of the paddlefish (*Polyodon spathula*). Journal of Morphology 140(2):243-256. (4,5)

546. WEISEL, G. F. 1975. The integument of the paddlefish, *Polyodon spathula*. Journal of Morphology 145(2):143-150. (4)

547. WHITE, C. E., JR. 1959. Selectivity and effectiveness of certain types of commercial nets in Tennessee Valley Authority lakes of Alabama. Transactions of the American Fisheries Society 88(2):81-87. (9)

548. WHITE, C. E., JR. 1961. The selectivity and effectiveness of bait and snag lines fished in the TVA lakes of Alabama. Proceedings of the Annual Conference Southeast Association of Game and Fish Commissioners. 15:359-367. (9)

549. WILDER, B. G. 1876. Notes on the North American ganoids; *Amia, Lepisosteus, Acipenser,* and *Polyodon*. Proceedings of the American Association for the Advancement of Science. 24th meeting (part 2):151-196. (6)

550. WILSON, W. D. 1956. Notes on cestodes in paddlefish, *Polyodon spathula* (Walbaum), from the Missouri River. Transactions of the Kansas Academy of Science 59(4):459-460. (15)

551. WOLLITZ, R. E. 1968. Smallmouth bass stream investigations. Clinch River study. Virginia Commission of Game and Inland Fisheries. Dingell-Johnson Project F-15-R-6, Job Number 3. Final Report. 93 pages. (1)

552. WU HSIEN WEN. 1930. On some fishes collected from the upper Yangtze Valley. Sinensia 1(6):65-86. (16)

553. YEAGER, B., AND R. WALLUS. 1982. Development of larval *Polyodon spathula* (Walbaum) from the Cumberland River, Tennessee. Pages 73-77 *in* C. F. Bryan, J. V. Connors, and F. M. Truesdale, editors. The proceedings of the fifth annual larval fish conference. Louisiana Cooperative Fishery

Research Unit, Louisiana State University, Baton Rouge, Louisiana, USA. 86 pages. (4,11,13)

554. ZEISBERGER, D. 1910. David Zeisberger's history of the Northern American Indians, edited by A. B. Hulbert and W. N. Schwarze. Ohio Archaeological and Historical Society Publication 19(1,2):1-189. (1,2,4)

555. ZOGRAFF, N. 1887. On some of the affinities between the ganoidei, chondrostei, and other fishes. Nature 37:70. (2)

Subject Headings

1. Distribution
2. Taxonomy
3. Evolution
4. Anatomy
5. Physiology
6. Natural History
7. Popular
8. Sport Fisheries
9. Commercial Fisheries
10. Management
11. Reproduction
12. Food Habits
13. Age and Growth
14. Culture
15. Parasites
16. Chinese Paddlefish (*Psephurus gladius*)

Subject Index

1. DISTRIBUTION

23	25	26	27	29	38	39	44	50	51
56	66	73	74	75	80	81	86	91	95
98	99	100	101	103	115	117	124	133	136
142	144	146	160	162	163	164	165	170	177
180	181	182	183	184	191	192	196	199	209
219	222	224	225	229	231	232	237	238	247
248	250	255	256	257	258	260	263	268	271
273	274	284	285	286	291	295	299	304	307
312	317	318	320	323	335	336	339	341	348
372	373	378	379	386	387	388	397	398	399
411	412	421	422	425	429	430	431	447	448
449	450	454	459	468	470	471	472	475	477
478	486	494	507	509	512	513	514	537	542
551	554								

2. TAXONOMY

6	21	51	165	219	229	250	269	270	272
285	286	293	294	304	305	317	322	342	412
414	430	458	459	507	534	537	554	555	

3. EVOLUTION

4	49	96	118	119	120	131	219	226	269
311	394	458							

4. ANATOMY

4	12	14	15	16	17	40	49	51	62
67	68	69	70	107	108	127	128	137	140
165	190	193	198	219	226	229	234	243	250
261	275	285	286	287	294	296	297	300	302
304	317	319	334	342	343	344	345	346	347
384	385	395	412	426	427	428	430	459	476
491	496	502	507	529	534	537	545	546	553
554									

5. PHYSIOLOGY

2	3	49	60	61	96	102	122	125	129
134	138	139	140	167	169	197	198	200	201
205	206	207	208	219	221	235	241	242	243
300	301	303	308	313	319	330	377	402	487
506	508	529	534	544	545				

6. NATURAL HISTORY

5	6	7	8	10	11	13	18	23	25
26	30	33	34	35	39	41	43	44	45
50	53	64	65	66	71	72	73	75	77
81	83	84	85	87	90	92	93	94	95
101	103	104	105	106	114	117	123	124	126
130	136	142	144	147	148	149	150	151	152
153	154	155	156	157	158	159	164	181	186
187	188	191	192	194	199	201	202	203	204
209	210	211	212	213	214	215	216	217	218
219	225	231	238	239	244	245	246	249	254
256	258	259	262	263	265	268	270	271	273
274	276	277	278	280	288	291	298	301	314
315	318	328	332	333	336	337	338	339	341
349	350	361	364	365	366	367	368	369	370
371	374	379	382	383	384	390	391	392	393
396	397	398	399	401	403	404	407	408	409
416	417	418	419	420	421	423	424	425	426
427	428	429	435	436	437	438	439	440	441
443	445	446	447	448	449	450	453	460	461
468	471	474	475	478	488	489	501	503	504
508	510	512	517	518	519	520	521	522	523
524	525	526	527	528	531	532	533	534	535
538	541	549							

7. POPULAR

1	20	22	24	33	37	41	43	53	54
64	65	76	83	84	94	121	132	148	168
173	185	204	211	219	224	249	260	265	279
290	310	321	327	351	387	388	389	401	405
406	415	416	417	424	433	435	445	473	474
485	503	525	532	533					

8. SPORT FISHERIES

1	18	19	20	24	42	46	47	48	54
55	84	95	109	110	111	112	113	114	147
148	149	150	151	152	153	154	155	156	157
158	159	174	175	185	186	187	188	189	194
195	196	210	212	213	214	215	216	217	218
219	227	228	236	245	276	277	278	290	321
349	350	351	353	354	355	356	357	358	359
360	361	362	363	364	365	366	367	368	369
370	371	374	375	376	392	403	404	405	408
409	417	418	419	420	421	423	426	428	429
434	436	437	438	439	440	441	443	447	448
449	450	453	462	463	464	465	473	479	481
482	483	484	488	517	518	519	520	521	522
523	524	526	527	528	530	531	534		

9. COMMERCIAL FISHERIES

7	9	18	19	32	58	71	72	76	78
79	85	92	95	97	105	106	109	110	111
112	113	116	145	161	166	181	194	195	196
219	239	245	259	262	264	266	267	283	301
325	326	328	338	390	393	396	451	457	469
489	497	499	500	511	530	534	536	547	548

10. MANAGEMENT

7	8	18	19	22	23	42	46	47	48
55	72	86	93	95	109	110	111	112	113
114	147	149	150	151	152	153	154	155	156
157	158	159	174	175	186	187	188	191	194
195	202	203	210	212	213	214	215	216	217
218	219	227	239	245	246	258	266	276	277
278	301	315	327	349	350	353	354	355	356
357	358	359	360	361	362	363	364	365	366
367	368	369	370	371	374	375	376	390	391
392	396	403	404	408	409	418	419	420	421
423	425	429	432	434	436	437	438	439	440
441	446	447	448	449	450	456	460	462	463
464	465	479	480	481	482	483	484	488	494
501	517	518	519	520	521	522	523	524	526
527	528	530	531	534	541				

11. REPRODUCTION

8	13	28	29	30	42	126	160	186	194
215	218	219	240	244	245	248	253	276	277
278	280	301	328	338	349	350	393	403	404
406	407	408	423	425	446	447	448	449	450
477	489	510	526	527	534	538	539	540	541
553									

12. FOOD HABITS

63	85	87	143	172	176	178	179	201	219
246	251	328	331	332	333	338	425	427	432
446	447	448	449	450	489	517	518	520	534

13. AGE AND GROWTH

5	7	19	42	45	52	55	59	63	72
75	87	93	109	110	111	112	113	186	187
188	194	195	201	202	203	215	218	219	245
252	253	254	301	307	324	328	331	338	392
393	403	404	408	409	418	419	423	425	429
443	446	447	448	449	450	460	490	495	505
517	518	521	522	524	527	534	541	553	

14. CULTURE

28	31	57	63	82	85	88	89	135	171
187	188	200	203	219	220	223	276	277	278
279	283	292	315	327	329	331	338	350	352
374	407	410	436	437	438	439	440	441	442
444	447	448	449	450	455	456	461	490	493
515	516	517	518	520	534	543			

15. PARASITES

36	219	306	309	328	340	413	452	466	467
492	534	550							

16. CHINESE PADDLEFISH (*Psephurus gladius*)

141	219	230	233	269	281	282	289	316	373
380	381	400	496	498	534	552			